# Pitman Research Notes in Mathematics Series

WITHDRAWN

**Main Editors**
H. Brezis, Université de Paris
R.G. Douglas, Texas A&M University
A. Jeffrey, University of Newcastle upon Tyne *(Founding Editor)*

**Editorial Board**
H. Amann, University of Zürich
R. Aris, University of Minnesota
G.I. Barenblatt, University of Cambridge
A. Bensoussan, INRIA, France
P. Bullen, University of British Columbia
S. Donaldson, University of Oxford
R.J. Elliott, University of Alberta
R.P. Gilbert, University of Delaware
R Glowinski, University of Houston
D. Jerison, Massachusetts Institute of Technology
K. Kirchgässner, Universität Stuttgart
B. Lawson, State University of New York at Stony Brook
B. Moodie, University of Alberta
S. Mori, Kyoto University
L.E. Payne, Cornell University
G.F. Roach, University of Strathclyde
I. Stakgold, University of Delaware
W.A. Strauss, Brown University
S.J. Taylor, University of Virginia

**Submission of proposals for consideration**
Suggestions for publication, in the form of outlines and representative samples, are invited by the Editorial Board for assessment. Intending authors should approach one of the main editors or another member of the Editorial Board, citing the relevant AMS subject classifications. Alternatively, outlines may be sent directly to the publisher's offices. Refereeing is by members of the board and other mathematical authorities in the topic concerned, throughout the world.

**Preparation of accepted manuscripts**
On acceptance of a proposal, the publisher will supply full instructions for the preparation of manuscripts in a form suitable for direct photo-lithographic reproduction. Specially printed grid sheets can be provided and a contribution is offered by the publisher towards the cost of typing. Word processor output, subject to the publisher's approval, is also acceptable.

Illustrations should be prepared by the authors, ready for direct reproduction without further improvement. The use of hand-drawn symbols should be avoided wherever possible, in order to maintain maximum clarity of the text.

The publisher will be pleased to give any guidance necessary during the preparation of a typescript, and will be happy to answer any queries.

**Important note**
In order to avoid later retyping, intending authors are strongly urged not to begin final preparation of a typescript before receiving the publisher's guidelines. In this way it is hoped to preserve the uniform appearance of the series.

**Addison Wesley Longman Ltd**
**Edinburgh Gate**
**Harlow, Essex, CM20 2JE**
**UK**
**(Telephone (0) 1279 623623)**

AF719289

**Titles in this series. A full list is available from the publisher on request.**

303 Numerical analysis 1993
**D F Griffiths and G A Watson**
304 Topics in abstract differential equations
**S Zaidman**
305 Complex analysis and its applications
**C C Yang, G C Wen, K Y Li and Y M Chiang**
306 Computational methods for fluid-structure interaction
**J M Crolet and R Ohayon**
307 Random geometrically graph directed self-similar multifractals
**L Olsen**
308 Progress in theoretical and computational fluid mechanics
**G P Galdi, J Málek and J Necas**
309 Variational methods in Lorentzian geometry
**A Masiello**
310 Stochastic analysis on infinite dimensional spaces
**H Kunita and H-H Kuo**
311 Representations of Lie groups and quantum groups
**V Baldoni and M Picardello**
312 Common zeros of polynomials in several variables and higher dimensional quadrature
**Y Xu**
313 Extending modules
**N V Dung, D van Huynh, P F Smith and R Wisbauer**
314 Progress in partial differential equations: the Metz surveys 3
**M Chipot, J Saint Jean Paulin and I Shafrir**
315 Refined large deviation limit theorems
**V Vinogradov**
316 Topological vector spaces, algebras and related areas
**A Lau and I Tweddle**
317 Integral methods in science and engineering
**C Constanda**
318 A method for computing unsteady flows in porous media
**R Raghavan and E Ozkan**
319 Asymptotic theories for plates and shells
**R P Gilbert and K Hackl**
320 Nonlinear variational problems and partial differential equations
**A Marino and M K V Murthy**
321 Topics in abstract differential equations II
**S Zaidman**
322 Diffraction by wedges
**B Budaev**
323 Free boundary problems: theory and applications
**J I Diaz, M A Herrero, A Liñan and J L Vazquez**
324 Recent developments in evolution equations
**A C McBride and G F Roach**
325 Elliptic and parabolic problems: Pont-à-Mousson 1994
**C Bandle, J Bemelmans, M Chipot, J Saint Jean Paulin and I Shafrir**
326 Calculus of variations, applications and computations: Pont-à-Mousson 1994
**C Bandle, J Bemelmans, M Chipot, J Saint Jean Paulin and I Shafrir**
327 Conjugate gradient type methods for ill-posed problems
**M Hanke**
328 A survey of preconditioned iterative methods
**A M Bruaset**
329 A generalized Taylor's formula for functions of several variables and certain of its applications
**J-A Riestra**
330 Semigroups of operators and spectral theory
**S Kantorovitz**
331 Boundary-field equation methods for a class of nonlinear problems
**G N Gatica and G C Hsiao**
332 Metrizable barrelled spaces
**J C Ferrando, M López Pellicer and L M Sánchez Ruiz**
333 Real and complex singularities
**W L Marar**
334 Hyperbolic sets, shadowing and persistence for noninvertible mappings in Banach spaces
**B Lani-Wayda**
335 Nonlinear dynamics and pattern formation in the natural environment
**A Doelman and A van Harten**
336 Developments in nonstandard mathematics
**N J Cutland, V Neves, F Oliveira and J Sousa-Pinto**
337 Topological circle planes and topological quadrangles
**A E Schroth**
338 Graph dynamics
**E Prisner**
339 Localization and sheaves: a relative point of view
**P Jara, A Verschoren and C Vidal**
340 Mathematical problems in semiconductor physics
**P Marcati, P A Markowich and R Natalini**
341 Surveying a dynamical system: a study of the Gray–Scott reaction in a two-phase reactor
**K Alhumaizi and R Aris**
342 Solution sets of differential equations in abstract spaces
**R Dragoni, J W Macki, P Nistri and P Zecca**
343 Nonlinear partial differential equations
**A Benkirane and J-P Gossez**
344 Numerical analysis 1995
**D F Griffiths and G A Watson**
345 Progress in partial differential equations: the Metz surveys 4
**M Chipot and I Shafrir**
346 Rings and radicals
**B J Gardner, Liu Shaoxue and R Wiegandt**
347 Complex analysis, harmonic analysis and applications
**R Deville, J Esterle, V Petkov, A Sebbar and A Yger**
348 The theory of quantaloids
**K I Rosenthal**
349 General theory of partial differential equations and microlocal analysis
**Qi Min-you and L Rodino**
350 Progress in elliptic and parabolic partial differential equations
**A Alvino, P Buonocore, V Ferone, E Giarrusso, S Matarasso, R Toscano and G Trombetti**

A Alvino, P Buonocore,
V Ferone, E Giarrusso
Università di Napoli "Federico II", Italy
S Matarasso
Università di Bologna, Italy
R Toscano
Seconda Università di Napoli, Italy
and
G Trombetti
Università di Napoli "Federico II", Italy
(Editors)

# Progress in elliptic and parabolic partial differential equations

LONGMAN

QA
370
.P767
1996

Addison Wesley Longman Limited
Edinburgh Gate, Harlow
Essex CM20 2JE, England
*and Associated Companies throughout the world.*

*Published in the United States of America
by Addison Wesley Longman Inc.*

© Addison Wesley Longman Limited 1996

All rights reserved; no part of this publication may be reproduced, stored in a retrieval system, or transmitted in any form or by any means, electronic, mechanical, photocopying, recording, or otherwise, without the prior written permission of the Publishers, or a licence permitting restricted copying in the United Kingdom issued by the Copyright Licensing Agency Ltd, 90 Tottenham Court Road, London, W1P 9HE.

First published 1996

AMS Subject Classifications: (Main) 35xx, 49xx
(Subsidiary) 46xx

ISSN 0269-3674

ISBN 0 582 25970 3

**British Library Cataloguing in Publication Data**

A catalogue record for this book is
available from the British Library

Printed and bound in Great Britain
by Biddles Ltd, Guildford and King's Lynn

## CONTENTS

# Preface

This book collects reports of the invited plenary addresses given during the conference "Elliptic and Parabolic P.D.E.'s and Applications", held in Capri, Italy, from September 19 to September 23, 1994.

The conference was devoted to new developments in Partial Differential Equations of Elliptic and Parabolic type, and to the applications in various fields.

The conference was attended by about 150 participants and among them many of the world's leading figures in the subject area were present. We would like to thank all the participants for their help in making the meeting successful and special thanks go to the contributors of this volume.

The following institutions, *Università di Napoli "Federico II", Seconda Università di Napoli, Dipartimento di Matematica e Applicazioni "R. Caccioppoli", Istituto Italiano per gli Studi Filosofici, Consiglio Nazionale delle Ricerche, Istituto per le Applicazioni della Matematica del C.N.R., Accademia di Scienze Fisiche e Matematiche - Napoli, C.I.R.A.M. - Università di Bologna, Regione Campania, Liguori Editore - Napoli, Banco di Napoli, Texas Instruments S.p.A.*, made possible the organization of the meeting with their financial support. We express our appreciation to them.

Finally, we thank Longman for helping us to publish these proceedings.

A. Alvino, P. Buonocore, V. Ferone, E. Giarrusso, S. Matarasso, R. Toscano, G. Trombetti

A. AMBROSETTI*

# Global branches of positive solutions of nonlinear eigenvalue problems

**1.** The existence of positive solutions of semilinear elliptic boundary value problems can be often studied by means of Bifurcation theoretical arguments jointly with suitable *a-priori* estimates.

Let $\Omega$ be a bounded domain in $\mathbb{R}^N$ and consider a nonlinear boundary value problem of the form

$$\begin{cases} -\Delta u &= \lambda f(x,u), \quad x \in \Omega, \\ u(x) &= 0, \qquad\qquad x \in \partial\Omega, \end{cases} \tag{1}$$

where $\lambda$ is a real parameter and $f$ is such that $f(x,0) \equiv 0$. Suppose that $f(x,\cdot)$ is differentiable at $u = 0$ and consider the linearized problem at $u = 0$

$$\begin{cases} -\Delta v &= \lambda f_u(x,0)v, \quad x \in \Omega, \\ v(x) &= 0, \qquad\qquad x \in \partial\Omega. \end{cases}$$

In the sequel let $\lambda_1[m,\Omega]$ ($m \in L^\infty(\Omega)$) denote the principal eigenvalue of

$$\begin{cases} -\Delta u &= \lambda m(x)u, \quad x \in \Omega, \\ u(x) &= 0, \qquad\qquad x \in \partial\Omega. \end{cases} \tag{2}$$

When $m \equiv 1$ the principal eigenvalue will be denoted by $\lambda_1[\Omega]$ or simply by $\lambda_1$. It is well known that the classical Rabinowitz *Global Bifurcation Theorem* [7] yields a global, unbounded branch of positive solutions $(\lambda, u)$ of (1) emanating from $(\lambda_1[f_u(x,0),\Omega],0)$. Moreover, $\lambda_1[f_u(x,0),\Omega]$ is the only bifurcation point of positive solutions of (1).

The global behaviour of such a branch obviously depends on the asymptotic properties of $f$. For example, when $f$ is asymptotically linear at infinity and $f(x,0) \geq 0$, a complete description of the set of positive solutions has been given in [5] and [6]. The case in which $f(x,0) < 0$ has been handled in [3]. Let us point out that in this latter case positive solutions bifurcate *from infinity*, not from the trivial solution $u = 0$.

---

* Supported by M.U.R.S.T., "Problemi Nonlineari...".

In a recent paper [4] we have also considered boundary value problems of the form

$$\begin{cases} -\Delta u &= \lambda u + g(x)h(u)u, \quad x \in \Omega, \\ u(x) &= 0, \qquad\qquad\qquad\quad x \in \partial\Omega, \end{cases} \tag{3}$$

where $h(0) = 0$, $h(u) \to -\infty$ as $u \to +\infty$, and $g$ can possibly vanish in a domain $\Omega_0 \subset \Omega$. Improving some results of [1], it is shown in [4] that the branch bifurcating from $(\lambda_1, 0)$ blows up at infinity if and only if $\lambda$ tends to $\lambda_0 := \lambda_0[\Omega_0]$ (when $\Omega_0$ is empty or has zero measure it is understood that $\lambda_0 = +\infty$). In particular, (3) has a positive solution for all $\lambda \in (\lambda_1, \lambda_0)$. Unlike [1] where a variational approach is emploied, we use bifurcation theory and suitable estimates. In the next section we will illustrate this procedure in more details, in the frame of a somewhat different problem discussed in [2].

**2.** Let $0 < q < p$ and consider the boundary value problem

$$\begin{cases} -\Delta u &= \lambda u + a(x)u^{q+1} - b(x)u^{p+1}, \quad x \in \Omega, \\ u(x) &= 0, \qquad\qquad\qquad\qquad\qquad\quad x \in \partial\Omega, \end{cases} \tag{4}$$

where $a, b \in L^\infty$ satisfy

$(A_1)$ there exists a smooth domain $\Omega_0 \subset \Omega$ such that $a(x) = b(x) = 0$ on $\Omega_0$ and $a(x) \geq 0$ and $b(x) > 0$ a.e. in $\Omega \setminus \Omega_0$.

According to the discussion in the preceding section, there exists a (connected) branch $\Sigma$ of positive solutions of (4) bifurcating from $(\lambda_1, 0)$. Hereafter, by a solution we mean a weak $C^{1,\nu}$ solution. Moreover $\Sigma$ is unbounded (say, in the topology of the uniform convergence). To control the behaviour of $\Sigma$ we set

$$\psi(x) = \begin{cases} \left(\dfrac{a^p}{b^q}\right)^{1/(p-q)} & \text{if } x \in \Omega \setminus \Omega_0 \\ 0 & \text{if } x \in \Omega_0, \end{cases}$$

and suppose

$(A_2)$ $\psi \in L^{N/2}(\Omega)$.

If $(\lambda, u)$ is a positive solution of (4) then there results

$$\int_\Omega |\nabla u|^2 dx = \lambda \int_\Omega u^2 dx + \int_\Omega (a(x)u^q - b(x)u^p)\, u^2 dx.$$

By elementary calculation one checks $\exists c_1 > 0$ such that

$$a(x)u^q - b(x)u^p \le c_1\psi(x) \qquad \text{a.e. in } \Omega\,. \tag{5}$$

Then one infers

$$\begin{aligned}\int_\Omega |\nabla u|^2 dx &\le \lambda\int_\Omega u^2 dx + c_1\int_\Omega \psi(x)u^2 dx \\ &\le \lambda\int_\Omega u^2 dx + c_1\|\psi\|_{L^{N/2}}\cdot\|u\|^2_{L^{2^*}}\end{aligned} \tag{6}$$

where $2^* = \dfrac{2N}{N-2}$. From (6) one deduces

$$\begin{aligned}c_2\|u\|^2_{L^{2^*}} &\le \int_\Omega |\nabla u|^2 dx \\ &\le \lambda c_3\|u\|^2_{L^{2^*}} + c_4\|u\|^2_{L^{2^*}}\end{aligned}$$

and this shows there exists $\lambda^* \in \mathbb{R}$ such that

$$\lambda \ge \lambda^* \tag{7}$$

for all positive solution $(\lambda, u)$ of (4).

**Remark 1** A direct application of the classical Theorem of *Bifurcation from the simple eigenvalue* allows us to say, in view of the specific form of the nonlinearity, that $\lambda^* < \lambda_1$, whenever $a \not\equiv 0$ a.e. in $\Omega \setminus \Omega_0$. ■

In order to prove the appropriate *a-priori* bound on positive solutions of (4), see Lemma 5 below, some preliminary results are in order.

Let $(\lambda_n, u_n)$ be a sequence of positive solutions of (4) such that

$$\|u_n\|_{L^2} \to +\infty.$$

Set

$$z_n = \frac{u_n}{\|u_n\|_{L^2}},$$

**Lemma 2** *If $\lambda_n \le \gamma < +\infty$ for some $\gamma > 0$, then there exists $C_1 > 0$ such that*

$$\int_\Omega |\nabla z_n|^2 dx \le C_1.$$

PROOF. From (4) and using (5) one finds

$$\begin{aligned}\int_\Omega |\nabla z_n|^2 dx &= \int_\Omega (\lambda_n + a(x)u_n^q - b(x)u_n^p)\, z_n^2 dx \\ &\le \int_\Omega (\gamma + c_1\psi(x))\, z_n^2 dx \le \gamma + c_1 \int_\Omega \psi(x) z_n^2\, dx. \qquad (8)\end{aligned}$$

As in [2] we set, for $\delta > 0$,

$$Y_\delta = \{x \in \Omega \setminus \Omega_0 \,:\, b(x) \le \delta\},$$

and note that $|Y_\delta| \to 0$ as $\delta \to 0$. Then, as a consequence of $(A_2)$, one deduces:

$$I_\delta := \left(\int_{Y_\delta} \psi^{N/2} dx\right)^{2/N} \to 0 \qquad (\delta \to 0). \qquad (9)$$

Moreover there results

$$\begin{aligned}\int_\Omega \psi z_n^2\, dx &= \int_{\Omega\setminus\Omega_0} \psi z_n^2\, dx \\ &\le \int_{Y_\delta} \psi z_n^2\, dx + \int_{(\Omega\setminus\Omega_0)\setminus Y_\delta} \psi z_n^2\, dx \\ &\le c(\delta)\|z_n\|_{L^2}^2 + c_2 I_\delta \cdot \|z_n\|_{L^{2^*}}^2 \\ &\le c(\delta) + c_3 I_\delta \cdot \int_\Omega |\nabla z_n|^2 dx, \qquad (10)\end{aligned}$$

where $c_3 > 0$ is a constant independent of $\delta$. According to (9) we can choose $\delta > 0$ in such a way that $c_1 \cdot c_3 \cdot I_\delta < 1$. Then, putting together (8) and (10), the Lemma follows. ■

The preceding Lemma implies:

**Corollary 3** *There exists $z \in W_0^{1,2}(\Omega)$, with $z \ge 0$ in $\Omega$ and $\|z\|_{L^2} = 1$, such that, up to a subsequence, $z_n \to z$ strongly in $L^2$.*

Next we show:

**Lemma 4** *$z(x) = 0$ a.e. in $\Omega \setminus \Omega_0$*

PROOF. Taken any $\varphi \in C_0^\infty(\Omega \setminus \Omega_0)$, one finds

$$\begin{aligned}-\int_{\Omega\setminus\Omega_0} z_n \Delta\varphi\, dx &= \lambda_n \int_{\Omega\setminus\Omega_0} z_n \varphi\, dx \\ &\quad + \int_{\Omega\setminus\Omega_0} (a(x)u_n^q - b(x)u_n^p)\, z_n \varphi\, dx.\end{aligned}$$

Since $z_n \to z$ a.e. in $\Omega$, we infer that

$$\liminf \int_{\Omega\setminus\Omega_0} (a(x)u_n^q - b(x)u_n^p)\, z_n\varphi\, dx > -\infty. \tag{11}$$

If $z(x) > 0$ in a set $D \subset \Omega \setminus \overline{\Omega}_0$ with $|D| > 0$, then

$$u_n(x) = z_n(x)\|u_n\|_{L^2} \to +\infty, \text{ a.e. in } D.$$

Since $b(x) > 0$ a.e. in $\Omega \setminus \Omega_0$ and $p > q$ it follows that

$$\int_D (a(x)u_n^q - b(x)u_n^p)\, z_n\varphi\, dx \to -\infty,$$

which is in contradiction with (11). ■

We are now ready to prove:

**Lemma 5** *For all $\gamma < \lambda_0$ there exists $C > 0$ such that*

$$\|u\|_{L^2} \le C,$$

*for all positive solution $(\lambda, u)$ of (4), with $\lambda \le \gamma$.*

PROOF. Arguing by contradiction, we assume there is a sequence of positive solutions $(\lambda_n, u_n)$ of (4) such that $\|u_n\|_{L^2} \to \infty$ and set $z_n = u_n\|u_n\|_{L^2}^{-1}$. By (7) we can also assume that $\lambda_n \to \overline{\lambda}$, for some $\overline{\lambda} \le \gamma < \lambda_0$. Moreover, according to Corollary 3 and Lemma 4, $z_n \to z$ in $L^2$ with $z = 0$ a.e. in $\Omega \setminus \Omega_0$ and thus $z \in W_0^{1,2}(\Omega_0)$. Since $a(x) = b(x) = 0$ in $\Omega_0$, it immediately follows that $z$ solves

$$\begin{cases} -\Delta z &= \overline{\lambda} z, \quad x \in \Omega_0, \\ z(x) &= 0, \quad x \in \partial\Omega_0. \end{cases}$$

Since $\|z\|_{L^2} = 1$ then $z > 0$ in $\Omega_0$ and we deduce that $\overline{\lambda} = \lambda_0$, a contradiction. ■

**Remark 6** The preceding proof shows that if $\|u_n\|_{L^2} \to +\infty$ then $\lambda_n \to \lambda_0$. In such a case (with the same notation introduced before) $z_n(x) \to z(x) = r\phi_0(x)$, $r > 0$, where $\phi_0$ denotes a positive eigenfunction associated to $\lambda_0$. In particular,

$$u_n(x) \to +\infty, \qquad \text{for a.e. } x \in \Omega_0. \tag{12}$$

■

We finally show

**Lemma 7** *Problem (4) has no positive solution for* $\lambda \geq \lambda_0$.

PROOF. Let $(\lambda, u)$ be any positive solution of (4). Since $a = b = 0$ in $\Omega_0$, $u$ satisfies $-\Delta u(x) = \lambda u(x)$, $\forall\, x \in \Omega_0$. On $\partial\Omega_0$ one has that $u > 0$ and $\partial\phi_0/\partial\mathbf{n} < 0$ ($\mathbf{n}$ denotes the unit outword pointing normal to $\partial\Omega_0$). Then there results

$$\lambda \int_{\Omega_0} u\phi_0 \, dx = -\int_{\Omega_0} \Delta u \cdot \phi_0 \, dx < -\int_{\Omega_0} u\Delta\phi_0 \, dx = \lambda_0 \int_{\Omega_0} u\phi_0 \, dx,$$

and the result follows. ∎

The preceding Lemmas, Remarks 1 and 6 and the bifurcation theoretical arguments discussed in section 1, readily yield:

**Theorem 8** *Suppose that* $(A_1 - A_2)$ *hold. Then (4) has a positive solution for all* $\lambda_1 < \lambda < \lambda_0$. *More precisely, there exists a connected branch* $\Sigma$ *of positive solutions of (4) bifurcating from* $(\lambda_1, 0)$ *which blows up as* $\lambda \to \lambda_0$, *in the sense that along* $\Sigma$ *one has*

$$\lim_{\lambda \to \lambda_0} \|u\|_{L^\infty} = +\infty.$$

*In addition, if* $a \not\equiv 0$ *a.e. in* $\Omega$, *there exists* $\underline{\lambda} < \lambda_1$ *such that*

*(i) (4) has a positive solution provided* $\underline{\lambda} \leq \lambda < \lambda_0$;

*(ii) (4) has at least two positive solutions provided* $\underline{\lambda} < \lambda < \lambda_1$.

**Remarks 9** *(i)* The Laplace operator can be substituted by any second order uniformly elliptic operator (not necessarily in a divergence form). In addition, the nonlinearity can be replaced by a more general one of the form $a(x)k(u)u - b(x)h(u)u$, where $k$ and $h$ have the same behaviour of $u^q$ and $u^p$, respectively.

*(ii)* If $\Omega_0 = \emptyset$, (4) possesses positive solutions for all $\lambda \geq \underline{\lambda}$.

*(iii)* If $a \equiv 0$ the preceding results are essentially those discussed in Section 2 of [4]. ∎

**3.** Bifurcation arguments can also be used to handle semilinear Schrödinger equations like

$$-\Delta u + V(x)u = \lambda u + h(x,u)u, \qquad x \in \mathbb{R}^N \tag{13}$$

for an appropriate class of nonlinearities $h$, with $h(x, 0) = 0$. For example, following [4], equation (13) can be approximated with Dirichlet boundary value problems $(P_R)$

on balls $B_R \subset \mathbb{R}^N$. Each $(P_R)$ has a connected branch $\Sigma_R$ of positive solutions bifurcating from $(\lambda_1[B_R], 0)$ and a branch of positive solutions of (13) is searched by letting $R \to +\infty$. This limiting procedure can be carried out under suitable assumptions on $V$ and $h$. Roughly, we have considered in [4] the case in which

(*i*) $h(x,u) \simeq -g(x)u^p$ with $p > 0$ and $g$ like in (3);

(*ii*) $\liminf_{|x| \to +\infty} V(x) = 0$ and the linear problem

$$-\Delta u + V(x)u = \lambda u, \quad x \in \mathbb{R}^N,$$

has a principal eigenvalue $\Lambda < 0$.

It turns out that, for all $\gamma < \min\{\lambda_0, 0\}$, the connected sets $\Sigma_{R,\gamma} = \{(\lambda, u) \in \Sigma_R : \lambda \leq \gamma\}$ *converge* (in an appropriate sense) to a connected branch $\Sigma$ bifurcating from $(\Lambda, 0)$ and consisting of positive solutions of (13), see [4] for more details and precise statements.

Suitable modifications of these arguments would permit to handle also Schrödinger equations (13) with nonlinearities of the form $h(x,u) = a(x)u^q - b(x)u^p$, like that discussed in section 2.

## References


[1] ALAMA S. - G. TARANTELLO, *On the solvability of a semilinear elliptic equation via an associated eigenvalue problem*. Math. Z., to appear.

[2] ALAMA S. - G. TARANTELLO, *Elliptic problems with nonlinearities indefinite in sign*. Preprint, 1994.

[3] AMBROSETTI A. - D. ARCOYA - B. BUFFONI , *Positive solutions for semipositone problems via bifurcation theory.* Diff. Integral Equat. **7**-3 (1994), 655-663.

[4] AMBROSETTI A. - J. GÁMEZ, *Branches of positive solutions for some semilinear Schrödinger equations*. Preprint S.N.S. Pisa, May 1995.

[5] AMBROSETTI A. - P. HESS, *Positive solutions of asymptotically linear elliptic eigenvalue problems*. J. Math. Anal. Appl. **73**-2 (1980), 411-422.

[6] HESS P. - T. KATO, *On some linear and nonlinear eigenvalue problems with an indefinite weight function*. Comm. in P.D.E. **5** (1980), 999-1030.

[7] Rabinowitz P.H., *Some global results for nonlinear eigenvalue problems.* J. Funct. Anal., **7** (1971), 487-513.

Antonio Ambrosetti

Scuola Normale Superiore,

Pisa, 56100 Italy.

C. BANDLE, M. FLUCHER

# Applications of harmonic transplantation to variational problems

## 1. Introduction

Various physical and geometrical quantities can be expressed in the form

$$Q_p(\Omega) \quad := \quad \min_{u \in H_0^1(\Omega)} \frac{\int_\Omega |\nabla u|^2}{\left(\int_\Omega |u|^p\right)^{2/p}} \tag{1}$$

where $\Omega$ is a domain in $I\!R^N$ and $H_0^1(\Omega)$ is the usual Sobolev space of functions vanishing at the boundary in the sense of the trace and having first order distributional derivatives in $L^2(\Omega)$. For instance $Q_2(\Omega) = \lambda_1(\Omega)$ is the principal Dirichlet eigenvalue. If $\Omega$ is a simply connected domain in the plane then $Q_1(\Omega) = 4/S(\Omega)$, where $S(\Omega)$ corresponds to the torsional rigidity of a cylindrical beam. Finally, if $1 \le p \le 2N/(N-2)$ then $Q_p(\Omega)$ is the best Sobolev constant.

## 2. Conformal transplantation

Important techniques for the construction of lower bounds for $Q_p(\Omega)$ are the different types of rearrangements and symmetrizations. An ingenious device to estimate $Q_p(\Omega)$ from above is the conformal transplantation introduced by Polya and Szegö [18]. They successfully applied it to eigenvalue and torsion problems. It consists of constructing a trial function for $Q_p(\Omega)$ in a simply connected domain $\Omega$ by means of the Riemann map $g : \Omega \to B_0^1$. To any function $u : B_0^1 \to I\!R$ it associates a function $u_x := u \circ g : \Omega \to I\!R$. If $g(x) = 0$, then the local amplification factor $r(x) := |g'(x)|^{-1}$ is called *conformal radius* of $\Omega$ at $x$. Important features of conformal transplantation are the invariance of the Dirichlet integral:

$$\int_\Omega |\nabla u_x|^2 \quad = \quad \int_{B_0^1} |\nabla u|^2 \tag{2}$$

and

$$\int_\Omega f \circ u_x \quad \ge \quad r^2(x) \int_{B_0^1} f \circ u \tag{3}$$

for every positive function $f : I\!R \to I\!R^+$ and radially symmetric $u : B_0^1 \to I\!R$. These two properties immediately imply that:

$$\lambda_1(\Omega) \leq \frac{j_{0,1}^2}{\max r^2}, \qquad S(\Omega) \geq \frac{\pi}{2\max r^4}.$$

Numerous extensions and generalizations of these inequalities can be found in Polya and Szegö [18], Szegö [19], Polya and Schiffer [17], Hersch [12, 13] and Bandle [4]. More recent applications to nonlinear problems were obtained by Flucher [8, 9, 10]. He mainly used conformal and harmonic transplantation to localize the points of concentration in variational problems with concentration of energy. A survey of these results is contained in [6].

## 3. Harmonic radius

First we extend the conformal radius to multiply connected and higher dimensional domains. To this end we divide the Green's function into a radially symmetric singular and a regular part:

$$G_x(y) = \gamma(F(|x-y|) - H_x(y))$$

given by

$$F(s) := \begin{cases} -\log(s) & (n=2), \\ s^{2-n} & (n \neq 2). \end{cases}$$

The regular part $H_x$ is a harmonic function with the same boundary values as the singular part. In particular we can evaluate it at the singularity.

**Definition 1 (Harmonic radius)** *The harmonic radius of $\Omega$ at $x$ is defined by*

$$r(x) := \begin{cases} e^{-H_x(x)}, & (n=2), \\ (H_x(x))^{\frac{1}{2-n}}, & (n \neq 2). \end{cases}$$

*The points in $\Omega$ where the harmonic radius assumes its maximum are called the harmonic centers of $\Omega$.*

For simply connected planar domains this definition coincides with the notion of inner or conformal radius. The harmonic radius is positive in the interior of $\Omega$ and vanishes at the boundary with slope 2 [6, 5]. In variational problems with concentration of energy the minimizers typically concentrate at the harmonic centers of the underlying domain. In particular the concentration point does not depend on the nonlinearity. See [6] for a list of such problems. Numerically the harmonic centers of a simply connected planar domain can be computed by solving Liouville's equation (Section 7).

## 4. Harmonic transplantation

Conformal transplantation is restricted to simply connected planar domains. A substitute for multiply connected and higher dimensionsional domains is the harmonic transplantation introduced by Hersch [13]. A careful examination of the proofs of (2) and (3) shows that these properties rely on the fact that $\log|g(x)|$ is harmonic. Since $G_x = -\frac{1}{2\pi}\log|g|$ is the Dirichlet Green's function of the Laplace operator on $\Omega$ with singularity at $x$, it seems natural to define a transplantation from a ball to an arbitrary domain in $\mathbb{R}^N$ by means of the Green's function.

**Definition 2 (Harmonic transplantation)** *If* $u : B_0^\rho \subset \mathbb{R}^N \to \mathbb{R}$ *is of the form* $\mu \circ G_0$*, then its harmonic transplantation to* $\Omega$ *centered at* $x$ *is given by* $u_x := \mu \circ G_x : \Omega \to \mathbb{R}$.

In contrast to the conformal transplantation the harmonic transplantation only applies to radially symmetric functions. On the other hand it shares the properties (2) and (3) of the conformal transplantation:

$$\int_\Omega |\nabla u_x|^2 = \int_{B_0^\rho} |\nabla u|^2, \tag{4}$$

$$\int_\Omega f(u_x) \geq \int_{B_0^{r(x)}} f(u). \tag{5}$$

The latter inequality will be proved in Section 6. Using harmonic tranplantation we can solve the following isoperimetric problem.

**Theorem 3.** *The maximum*

$$\sup\left\{Q_p(\Omega) \ : \ \Omega \subset \mathbb{R}^N, \ \max r \geq R\right\}$$

*is attained for balls of radius* $R$.

By means of symmetrization it is easily shown that among all domains of given volume the balls yield the minimum of $Q_p(\Omega)$. By means of the Carleman-Szegö inequality for the capacity it can be shown that

$$|\Omega| \geq \max r^N \left|B_0^1\right|.$$

This implies that among all domains with $\max r \geq R$ the balls have the least volume. Using complex function theory, Polya and Szegö proved that in the plane the solution of the torsion problem

$$\begin{aligned} -\Delta u &= 1 \text{ in } \Omega, \\ u &= 0 \text{ on } \partial\Omega \end{aligned}$$

satisfies $u(x) \geq r^2(x)/2N$. This result was generalized by Payne to arbitrary dimensions. It can be easily derived by means of harmonic transplantation.

## 5. Harmonic transplantation by capacity potential

In this section we describe a generalization of the harmonic transplantation. It was already used by Hersch [13]. Consider a subset $A \subset \Omega$ with smooth boundary and let $k$ be its capacity potential defined by:

$$\begin{aligned} -\Delta k &= 0 \text{ in } \Omega \setminus A, \\ k &= 1 \text{ on } A, \\ k &= 0 \text{ on } \partial\Omega. \end{aligned}$$

The quantity

$$cap_\Omega(A) := \int_{\Omega\setminus A} |\nabla k|^2 = \int_{\partial A} |\nabla k| \tag{6}$$

is called the *capacity* of $A$ with respect to $\Omega$. By the maximum principle $k$ assumes values between 0 and 1. Set:

$$\begin{aligned} \Omega(t) &:= \{x \in \Omega \, : \, k(x) > t\}, \\ m(t) &:= |\Omega(t)|. \end{aligned}$$

The function $m$ is called distribution function of $k$. Harmonic transplantation by the capacity potential $k$ is defined as follows. Let $v : \Omega \setminus A \to I\!R$ be any function in $\Omega \setminus A$ with the same level sets as $k$. It can be written as $v = \nu \circ k$ for a suitable function $\nu$. Applying the divergence theorem and the coarea formula we get for the Dirichlet integral of $v$:

$$\int_{\Omega\setminus A} |\nabla v|^2 = cap_\Omega(A) \int_0^1 (\nu')^2.$$

Furthermore we have for any $f$:

$$\int_{\Omega\setminus A} f(v) \;=\; -\int_{\min v}^{\max v} f(\nu(t))dm(t).$$

The harmonic transplantation by means of the Green's function is a special case of harmonic transplantation by capacity potential. Indeed it suffices to take $A = \{G_x > t\}$ and to let $t$ tend to $\infty$. In this case $\mathrm{cap}_\Omega(A) = t$ and $k = G_x/t$. We now solve the following isoperimetric problem.

**Theorem 4.** *The maximum*

$$\sup\Big\{Q_p(\Omega\setminus A)\ :\ \Omega\subset I\!R^N,\ A\subset\Omega,\ |A|=\alpha,\ \mathrm{cap}_\Omega(A)=c\Big\}$$

*is attained for annuli* $\Omega\setminus A = B_x^R\setminus B_x^\rho$.

*Proof:* Chose $\Omega$ and $A$ as above with capacity potential $k$ and distribution function $m$. By the coarea formula

$$-\frac{dm(t)}{dt} \;=\; \int_{\{v=t\}} |\nabla k|^{-1}. \tag{7}$$

By (6), (7) and the Schwarz inequality

$$-\frac{dm(t)}{dt} \;\geq\; \frac{L^2(t)}{cap_\Omega(A)} \tag{8}$$

where $L(t) := |\{v=t\}|$. Applying the isoperimetric inequality $L(t)\geq \omega_N m^{(N-1)/N}(t)$ we get

$$-\,dm(t) \;\geq\; \frac{\omega_N^2}{cap_\Omega(A)} m^{\frac{2(N-1)}{N}}(t)dt. \tag{9}$$

Integration yields for $N>2$ the pair of inequalities

$$\begin{aligned}
&\frac{b_N}{cap_\Omega(A)}t + \frac{1}{m^{(N-2)/N}(0)} \\
\leq\;& \frac{1}{m^{(N-2)/N}(t)} \\
\leq\;& \frac{b_N}{cap_\Omega(A)}(t-1) + \frac{1}{m^{(N-2)/N}(1)}.
\end{aligned} \tag{10}$$

Equality holds if $\Omega(t)$ is a ball. A similar result can be derived in the plane. Rather than concentrating on the the first inequality which leads to the familiar results obtained by means of symmetrization techniques, we focus on the second one. Consider the concentric balls $A_0 \subset \Omega_0$ defined by $|A_0| = \alpha$ and $\text{cap}_{\Omega_0}(A_0) = c$. By the inequality of Carleman and Szegö $\Omega_0 \subset \Omega^*$. In the sequel all quantities related to $\Omega_0 \setminus A_0$ are denoted with an index 0. From (10) it follows that

$$m(t) \;\geq\; (\alpha_N(\Omega) + \beta_N(\Omega)t)^{-\frac{N}{N-2}}$$

and thus

$$m(t) \;\geq\; m_0(t).$$

Inserting this inequality into (9) we get

$$-\frac{dm(t)}{dt} \;\geq\; -\frac{dm_0(t)}{dt}.$$

This implies:

$$\int_{\Omega \setminus A} f \circ k \;\geq\; \int_{\Omega_0 \setminus A_0} f \circ k_0.$$

for any positive function $f : I\!R^+ \to I\!R^+$. In particular this holds for $f(t) = |t|^p$ which yields the desired result.

The special case of the principal eigenvalue of a multiply connected region ($p = 2$) was treated by Gasser and Hersch [11].

## 6. Lower bound for the distribution function of the Green's function

Inequality (4) follows from an estimate of the distribution function $m$ of $G_x$. By the same arguments as in the proof of Theorem 4 we derive a lower bound for $m$. The coarea formula yields

$$-\frac{dm(t)}{dt} \;=\; \int_{\partial\{G_x > t\}} |\nabla G_x|^{-1}. \tag{11}$$

Observe that

$$\int_{\partial\{G_x > t\}} |\nabla G_x| \;=\; 1. \tag{12}$$

From (11), (12) and Schwarz's inequality we have as in (8)

$$-\frac{dm(t)}{dt} \;\geq\; L^2(t).$$

By virtue of the isoperimetric inequality we get as in (9)

$$-\frac{dm(t)}{dt} \geq \omega_N^2 m^{2(N-1)/N}(t).$$

If we integrate this inequality from t to $\infty$ and make use of the expansion of the Green's function near the singularity: $G_x(y) = \gamma\,(F(|x-y| - F(r(x)) + O(|x-y|))$, we find

$$m(t) \geq m_0(t)$$

where $m_0$ is the distribution function of $G_0$ on $B_0^{r(x)}$.

## 7. Conformal radius and Liouville's equation

It was already observed by Liouville [14] in 1853 that the solutions of the equation

$$\begin{aligned} \Delta U &= 4e^{2U}, \\ U &= \infty \text{ on } \partial\Omega \end{aligned}$$

can be characterized by means of analytic functions. In fact

$$U = -\log(r) = -\log\left(\frac{1-|g|^2}{|g'|}\right)$$

with $g$ as in Section 1. The solution of Liouville's equation is unique. It is the maximal solution. It can be shown that $U(x) + \log(2\,dist(x, \partial\Omega)) \to 0$ as $x \to \partial\Omega$ [5]. This equivalent to the fact that the slope of the conformal radius at the boundary is 2. The conformal centers are the points where $U$ assumes its minimum. The harmonic radius of a multiply conncted domain cannot be expressd in terms of the maximal solution of the Liouville equation. For further discussion on this topic see [6].

## 8. Extension to manifolds

Let $M$ be a 2-dimensional manifold and let $\Omega_M$ be a domain on $M$ represented by $\Omega \subset I\!R^2$. Let $ds := p(x)|dx|$ be the Riemannian metric on $M$ given in isothermal coordinates. On sufficiently regular two-manifolds it is always possible to introduce such coordinates. The Green's function of $\Omega_M$ is the same as the Green's function of

$\Omega$ in $\mathbb{R}^2$. Let $r$ be the harmonic radius of $\Omega$. Then $r_M(x) := p(x)r(x)$ is the harmonic radius of $\Omega_M$. If $\Omega_M$ is simply connected, $U := -\log(r_M)$ satisfies the equation

$$\Delta U = 4p^2e^{-2U} - \Delta\log(p).$$

Since $-\Delta\log(p) = K\,p^2$ where $K$ denotes the Gaussian curvature of $M$, we get

$$p^{-2}\Delta U = 4e^{-2U} + K. \tag{13}$$

The operator on the left hand side of (13) is the Laplace-Beltrami operator on $M$. By means of the isoperimetric inequalities of Alexandrov (cf. [4]), it is possible to estimate the distribution function of the Green's function with respect to the metric $ds$:

$$m_M(t) = \int_{\Omega(t)} ds = \int_{\Omega(t)} p^2(x)dx.$$

It turns out that if $K \leq K_0$ for some constant $K_0$, then for any domain $\Omega_M$ the isoprimetric inequality $|\partial\Omega_M|^2 \geq (4\pi - K_0\,|\Omega_M|)\,|\Omega_M|$ holds. Similarly as in Section 6 we find

$$-\frac{dm_M}{dt} \geq (4\pi - m_M)m_M.$$

Hence

$$\frac{4\,m_M(t)}{4\pi - K_0 m_M(t)} \geq r_M^2(x)e^{-4\pi t}.$$

This allows to extend the harmonic transplantation to manifolds. Many isoperimetric inequalities for eigenvalues and for linear and nonlinear boundary value problems can be obtained in this way [1, 2, 3, 4]. Recently the authors have also investigated the higher dimensional case [7]. The methods described here extend to conformally flat spaces of constant curvature.

## References


[1] Ahlfors L.V., *Geodesic curvature and area*, Studies in Mathematical Analysis and Related Topics, Essays in Honor of G.Polya. Stanford University Press (1962), 1-7.

[2] Bandle C., *Extensions of an inequality by Polya and Schiffer for vibrating membranes*, *Pac. J. Math.*. (1972), 543-555.

[3] BANDLE C., *Mean value theorems for functions satisfying the inequality* $\Delta u + Ke^u \geq 0$, *Arch. Rat. Mech. Anal.*. (1973), 70-84.

[4] BANDLE C., *Isoperimetric Inequalities and Applications*, Pitman (1980).

[5] BANDLE C., ESSÉN M., *On the solutions of quasi-linear elliptic problems with boundary blow-up*, *Symposia Math.*. (Cambrigde 1994), 93-111.

[6] BANDLE C., FLUCHER M., *Harmonic radius and concentration of energy, hyperbolic radius and Liouville's equations* $\Delta u = e^u$ *and* $\Delta u = u^{\frac{n+2}{n-2}}$, To appear in SIAM Rev..

[7] BANDLE C., FLUCHER M., *Harmonic radius of manifolds.*, In preparation..

[8] FLUCHER M., *Extremal functions for the Trudinger-Moser inequality in two dimensions*, *Comment. Math. Helv.*. (1992), 471-497.

[9] FLUCHER M., *An asymptotic formula for the minimal capacity among sets of equal area*, **Calc. Var.**. (1993), 71-86.

[10] FLUCHER M., MÜLLER S., *Concentration of low energy extremals, identification of concentration points*, In preparation..

[11] GASSER T., HERSCH J., *Über die Eigenwerte einer mehrfach zusammenhängenden Membran: Erweiterung von isoperimetrischen Sätzen von Polya und Szegö*, *ZAMP.* (1968), 672-675.

[12] HERSCH J., *On symmetric membranes and conformal radius: some complements to Pólya's and Szegö's inequalities*, Arch. Rat. Mech. Anal..s (1965), 378-390.

[13] HERSCH J., *Transplantation harmonique, transplantation par modules, et théorèmes isopérimétriques*, *Comment. Math. Helvetici.* (1969), 354-366.

[14] LIOUVILLE J., *Sur l'équation aux différences partielles* $\partial^2 \log \lambda / \partial u \partial v \pm \lambda / 2a^2 = 0$, *J. de Math.*. (1853), 71-72.

[15] PAYNE L.E., *Torsion problem for multiply connected regions*, Studies in Mathematical Analysis and Related Topic, Essays in honor of G. Pólya. Stanford University Press (1962), 270-280.

[16] PAYNE L.E., *Isoperimetric inequalities and their applications*, *SIAM Rev.*. (1967), 453-488.

[17] PÓLYA G., SCHIFFER M., *Convexity of functionals by transplantation*, J. Analyse *Math.*. (1954), 245-345.

[18] PÓLYA G., SZEGÖ G., *Isoperimetric inequalities in mathematical physics*, Princeton University Press 1951.

[19] SZEGÖ G., *Inequalities for certain eigenvalues of a membrane of given area*, J. Rat. *Anal. Mech.*. (1954), 343-356.

[20] WEINBERGER H., *Symmetrization in uniformly elliptic problems*, Studies in Mathematical Analysis and Related Topics: Essays in honor of G.Polya, Stanford University Press, Stanford (1962), 424-428.

Catherine BANDLE, Martin FLUCHER

Universität Basel, Mathematisches Institut,

Rheinsprung 21, CH - 4051 Basel, Switzerland

bandle@math.unibas.ch, flucher@math.unibas.ch

H. BERESTYCKI

# Qualitative properties of semilinear elliptic equations in unbounded domains

It is well known that positive solutions of semilinear elliptic problems in a bounded domain have symmetry properties which reflect the symmetries of the operator and of the domain. Consider the following boundary value problem

$$
(1) \qquad \begin{cases} \Delta u + f(u) = 0 & \text{in } \Omega \\ u > 0 & \text{in } \Omega \\ u = 0 & \text{on } \partial\Omega. \end{cases}
$$

Here and in the following $f$ will always be assumed to be locally Lipschitz continuous: $\mathbb{R}^+ \to \mathbb{R}$.

A classical result of Gidas, Ni and Nirenberg [GNN1] asserts that if $\Omega$ is a sphere, say

$$\Omega = B_R(0) = \{x \in \mathbb{R}^n, |x| < R\}$$

then any solution of (1) has spherical symmetry and $\frac{\partial u}{\partial r} < 0$ for $0 < r = |x| < R$. This result was established by the *moving planes method* first devised by A.D. Alexandrov [A] in the study of minimal surfaces and then used by J. Serrin [S] for a symmetry result for over-determined problems which will be recalled in the last section of this paper. A more general version of the previous result is stated when $\Omega$ is any open set, convex in some direction $\xi$, and symmetric with respect to the hyperplane $\xi^\perp$. Then, $u$ is symmetric with respect to this hyperplane and $\dfrac{\partial u}{\partial \xi} < 0$ in $\{x \in \Omega; x \cdot \xi > 0\}$. In this generality (with no assumption on the smoothness of the boundary), this was proved in [BN1] where the method of moving planes and that of sliding domains —which is related— was simplified and made more general by the use of a version of the maximum principle in domains of small volume. I will describe this method in section 3 where it is used to prove a monotonicity result in an unbounded domain (Theorem 7). The maximum principle in domains of small volume is stated in Theorem 8 below.

In the paper [BN1] much more general equations than (1) — fully nonlinear elliptic equations: $F(x, u, Du, D^2u) = 0$ — are considered for which a new method involving an auxiliary evolution problem is developed. Also, other type of boundary conditions (e.g. mixed boundary conditions), as well as monotonicity results are treated in [BN2] and [BN3].

For the sake of simplicity, we restrict our attention here to the case of the equation (1) where $f$ only depends on $u$. In several results which will be described here, though, one may also allow some dependence of $f$ on $x$ and on $Du$.

Recently, several results about the qualitative properties of elliptic equations have been obtained for *unbounded domains*. Even though much progress has been achieved, the same generality as that available for bounded domains —in the results which I have recalled— has not yet been achieved for unbounded domains. There are many open problems with a varied range of difficulty and there are many interesting conjectures. I will indicate some of them along this paper.

This talk is intended as a partial survey of the subject. I will not discuss here a particularly important case of unbounded domain, that of the whole space for which there is a large literature. Most of the results which are reported here are taken from joint work with Luis Caffarelli and Louis Nirenberg (in particular [BCN2-6]) and are part of a program in understanding the structure of this type of problems.

In particular, I will not discuss here the important case of the whole space, when $\Omega = \mathbb{R}^n$. There is a vast literature on this problem. One may find general symmetry results in all of space in the works of Gidas, Ni and Nirenberg [GNN2] and Congming Li [L]. There is an interesting monotonicity result in a dimension two related to a conjecture of De Giorgi in Modica and Mortola [MM].

## 1. The case of a half space

The first case which I describe is that of a half space

$$\Omega = \{x \in \mathbb{R}^n; x = (x_1, ..., x_n), x_n > 0\}.$$

Here we consider *bounded* solutions of (1) in $\Omega$ but we do not assume a priori any conditions on the behaviour of $u(x)$ as $|x| \to \infty$ in $\Omega$. The type of properties that we are looking for here is that of symmetry and monotonicity. Symmetry in this context means that $u(x)$ is actually a function of $x_n$ alone and monotonicity refers to the property that $\frac{\partial u}{\partial x_n} > 0$ in $\Omega$. These properties are indeed the exact analogues in the case of a half space of the symmetry and monotonicity properties derived by Gidas, Ni and Nirenberg in a finite sphere.

One can think of the half space as an infinite sphere whose center goes off to infinity. (Indeed, $\Omega$ is the limit of the sphere $S_R = \{x \in \mathbb{R}^n, |x - Re_n| < R\}$ as $R \to \infty$).

Throughout the paper it will be assume that $f$ is Lipschitz continuous. In [BCN2-3], we have obtained several results of symmetry and monotonicity for the half space under various additional assumptions.

The first one involves an assumption on $f(\sup_\Omega u)$.

THEOREM 1. *Suppose $\Omega$ is a half space $\Omega = \{x \in \mathbb{R}^n; x_n > 0\}$ and let $u$ be a bounded positive solution of (1). Denote $M = \sup_\Omega u$ and assume that $f(M) \leq 0$. Then, necessarily, $u$ is a function of $x_n$ alone : $u = u(x_n)$ and $u'(x_n) > 0$ in $\Omega$. Moreover, $u(x) \nearrow M$ as $x_n \nearrow \infty$ and $f(M) = 0$.*

In this result, the assumption that $u$ is bounded is crucial. Indeed, for instance in the half plane $\Omega = \{x = (x_1, x_2) \in \mathbb{R}^2, x_2 > 0\}$, the function $u(x_1, x_2) = x_2 e^{x_1}$ satisfies $\Delta u - u = 0$ in $\Omega$ and is positive. Nervertheless, $u$ genuinely depends on $x_1$. Notice that in this example $M = +\infty$ and $f(+\infty) = -\infty \leq 0$. Theorem 1 is proved in [BCN2]. There, we also prove more general versions, in particular, when $f$ also depends on $x_n$.

Consider the problem

$$
(2) \qquad \begin{cases} \Delta u + \beta(x_n, u) = 0 & \text{in } \Omega = \{x_n > 0\}, \\ 0 < u \leq \sup_\Omega u =: M & \text{in } \Omega, \\ u = 0 & \text{on } \partial\Omega. \end{cases}
$$

Here, we assume that $\beta = \beta(x_n, t) : \mathbb{R}^+ \times \mathbb{R} \to \mathbb{R}$ is continuous and Lipschitz continuous in $t$ uniformly with respect to $x_n$.

THEOREM 2. *Suppose that $x_n \mapsto \beta(x_n, t)$ is nonincreasing and that $\beta(x_n, M) \leq 0$*

*for all $x_n \in \mathbb{R}_+$. Then a solution $u$ of (2) is symmetric increasing in the sense of Theorem 1.*

Another type of results is obtained under an assumption on the sign of $f(0)$. But before indicating it, let me recall an earlier result of *monotonocity* due to Dancer [D].

THEOREM (DANCER). *Assume that $f(0) \geq 0$. Then any bounded soluton $u$ of (1) in a half space $\Omega = \{x_n > 0\}$ is monotonic in the sense that*

$$\frac{\partial u}{\partial x_n} > 0 \text{ in } \Omega.$$

Actually Dancer assumes in [D] that either $f(0) > 0$ or $f(0) = 0$ and $f'(0) \geq 0$. But a slight modification of his proof yields the result under the more general assumption $f(0) \geq 0$.

Notice that Dancer's result only asserts a monotonicity property and does not say anything about symmetry.

I counter-example to Theorem 1 above with an unbounded $u$, the function is still monotonic with respect to $x_n$. Hence, there is a natural question which arises here :

*Open problem 1.* Does Dancer's result hold in general for unbounded solutions $u$ as well?

Starting with this monotonicity property, one can further derive *symmetry* in low dimensions.

THEOREM 3. *Suppose that $f(0) \geq 0$ and that the dimension $n$ is either 2 or 3. Then, any bounded solution $u$ of (1) in a half space $\Omega = \{x_n > 0\}$ is symmetric and monotonic.*

This result is proved in [BCN3] by means of a somewhat surprising link with a problem concerning Schrödinger operators that I state now.

In the whole space $\mathbb{R}^m$ consider a bounded potential $V(x) \in L^\infty(\mathbb{R}^m)$ and consider the Schrödinger operator

$$L = -\Delta + V(x).$$

Here is our result.

THEOREM 4. *Suppose that there exists a generalized eigenfunction $\psi$ associated to the "eigenvalue" 0 of the operator $L$ in the sense that*

$$\psi \in L^\infty(\mathbb{R}^m) \quad \text{and} \quad L\psi = 0.$$

*Assume further that $\psi$ changes sign. Then, in dimension $m = 1$ and $m = 2$, the operator $L$ has negative spectrum, that is, there exists a test function $\varphi \in C_0^\infty(\mathbb{R}^m)$ such that*

$$\int_{\mathbb{R}^m} \{|\nabla\varphi(x)|^2 + V(x)|\varphi(x)|^2\}dx \ < \ 0.$$

In fact, in [BCN3], we have reduced Theorem 3 in any dimension to the Schrödinger operator result of Theorem 4. The restriction on the dimension $n \leq 3$ corresponds to the restriction $m = 1, 2$ in Theorem 4. Were Theorem 4 to be proved in higher dimensions as well, then Theorem 3 would follow automatically via the reduction in [BCN3].

This property about Schrödinger operators is a very natural question which does not seem to have been considered in the specialized literature so far. So it is natural to pose the following question.

*Open problem 2.* Does the property for Schrödinger operators in Theorem 4 hold in higher dimensions as well?

Hence, in dimension $n = 2$ and 3, the only case to be treated would be that of an $f$ with $f(0) < 0$ and $f(M) > 0$. It is tempting to conjecture that there are no solutions in this case. A typical example is that of the following problem

$$\begin{cases} \Delta u + u - 1 = 0 \text{ in } \Omega = \{x_n > 0\} \\ 0 < u \leq M = \sup u < \infty \text{ in } \Omega \\ u = 0 \text{ on } \partial\Omega. \end{cases} \tag{3}$$

For this particular case, we were able to prove non existence in low dimensions.

PROPOSITION 5. *Assume that $n = 2$ or 3. Then, there is no solution of (3).*

This brings one naturally to the question of what happens for larger $n$.

*Open problem 3.* Are there solutions of (3) for large $n$?

The results in Theorem 1 and Theorem 3 yield symmetry properties for two seemingly unrelated general classes of problems. It is natural to ask whether one can prove the symmetry property in general.

*Open problem 4.* Assume that $f$ is Lipschitz continuous. Prove that any bounded solution of (1) in a half space is symmetric.

I conjecture that, indeed, the result holds without any further condition whatsoever.

If Theorem 3 gives a clear condition for symmetry, one may wonder about the nature of the condition $f(M) \leq 0$ in Theorem 1. In fact, Theorem 1 is a partial version of the full theorem. Indeed, if the general result were to be true, then necessarily any solution of (1) would satisfy $f(M) = 0$ and, thus, would fall within the statement of Theorem 1...

## 2. The case of a Lipschitz graph

The next setting which I consider now extends that of the half space: $\Omega$ is a domain bounded by a Lipschit graph. More precisely, let

$$\varphi : \mathbb{R}^{n-1} \to \mathbb{R}$$

denote a global Lipschitz function and let us assume here that $\Omega$ is given by

$$\Omega = \{x = (x_1, ..., x_n);\ x_n > \varphi(x_1, ..., x_{n-1})\}.$$

The solutions of (1) are always assumed to be of class $C^2$ in $\Omega$ and continuous up to the boundary.

The aim here is to achieve a *monotonicity* result in the $x_n$ direction. The function $f$ will be required to satisfy the following assumptions

$$f > 0 \text{ in } (0, \mu); \quad f(0) = f(\mu) = 0 \tag{4}$$

(5) $$f(s) \geq \eta s \text{ for } s \in [0, \delta] \text{ for some } \eta > 0 \text{ and } \delta > 0$$

(6) $$f(s) \text{ is nonincreasing for } s \in [\mu - \delta, \mu], \quad \delta > 0$$

(7) $$f \text{ is Lipschitz continuous on } [0, \mu].$$

THEOREM 6. *Under the above conditions (4)-(7), let $u$ be a solution of (1) such that $M := \sup_{\Omega} u \leq \mu$. Then, $u$ satisfies*

$$\frac{\partial u}{\partial x_n} > 0 \text{ in } \Sigma.$$

This Theorem is proved in our work with Caffarelli and Nirenberg [BCN4]. It rests on an adaptation of the *sliding method* (see [BN1]). In contrast, for the monotonicity result of Dancer in a half space which is recalled above, he uses a method of *reflection in moving planes.*

The original motivation for our result in [BCN4] stemmed from the study of regularization of some free boundary problems. Indeed, in certain settings, a free boundary problem may be approximated ("regularized") by a semi-linear elliptic problem of the type (1). I refer in particular to our work [BCN1] where we study this type of questions. The study of regularity often involves a *blow up* version of the problem. Roughly speaking, if one starts with a free boundary which is Lipschitz continuous, the blown up version of the approximated problem will be set precisely in a domain of the type considered here. (The blow up limit of a Lipschitz graph is a global Lipschitz graph). The monotonicity result in Theorem 6 then provides some information which can be translated in terms of uniform regularity for the level sets approximating the free boundary problems. Let me emphasize that the assumptions (4)-(7) allow one to treat precisely the problem arising in combustion theory which is considered in [BCN1].

There have been some earlier related works on symmetry and monotonicity properties.

The first result of this kind to my knowledge is due to M. Esteban and P.-L. Lions [EL]. They consider the case of what might be called a *coercive* Lipschitz graph, that is, they assume that

$$\lim_{\substack{|x|\to\infty \\ x\in\mathbb{R}^{n-1}}} \varphi(x) = +\infty. \tag{8}$$

They prove a weaker version of the following

THEOREM 7. *If $\Omega$ is bounded by a coercive continuous graph (in the above understanding), then, for any locally Lipschitz function $f$ and given any solution $u$ of (1) –be it bounded or not–, it satisfies*

$$\frac{\partial u}{\partial x_n} > 0 \text{ in } \Omega.$$

Notice that for this result the graph $\varphi$ needs not be globally Lipschitz. Actually, Esteban and Lions get this result for a smooth graph $\varphi$. But using the improved version of the moving plane method which we developed with L. Nirenberg in [BN1], it is possible to extend this result to the case of a merely continuous graph $\varphi$.

This result rests on the method of reflection in moving planes.

The observation of M. Esteban and P.-L. Lions [EL] is that this method, used for *bounded* domains, readily extends to the unbounded case when the coercivity condition (8) is met.

Since it is simple, let me sketch the argument. This will give me the occasion to present the approach to the moving planes method of Nirenberg and myself [BN1]. It is both simpler than the original method and gives more general results.

I will use some classical notations. For $\lambda \in \mathbb{R}$, $T_\lambda = \{x_n = \lambda\}$ is the *moving hyperplane*, $\lambda_0 = \inf\{\lambda \in \mathbb{R}; T_\lambda \cap \Omega \neq \emptyset\}$, (this number $\lambda_0$ is finite because of (8))

$$\Omega_\lambda = \{x \in \Omega_\lambda; x_n < \lambda\}$$

is the lower cap of $\Omega$ cut out by $T_\lambda$.

For $x \in \Omega_\lambda$, $x = (x_1, ..., x_n)$, let $x^\lambda = (x_1, ..., x_{n-1}, 2\lambda - x_n)$ denote its reflection in $\Omega$,

$$v^\lambda(x) = u(x^\lambda)$$
$$w^\lambda(x) = u(x^\lambda) - u(x).$$

These two functions are defined in $\Omega_\lambda$.

Notice that since $\Omega$ is defined as an epigraph, the cap $\Omega_\lambda$ is always included in $\Omega$ and is not empty for all $\lambda > \lambda_0$. The main consequence of the coercivity assumption (8) is that $\Omega_\lambda$ is always *bounded*, for all values of $\lambda$. Actually, (8) is equivalent to the boundedness of $\Omega_\lambda$ for all $\lambda$. This is the reason why the moving planes method for bounded domains applies as such to epigraphs when (8) is satisfied.

Indeed, as usual, essentially, one wants to prove that $w^\lambda > 0$ in $\Omega_\lambda$ for all $\lambda > \lambda_0$. For this shows that $u$ is nondecreasing with respect to $x_n$, hence that $\frac{\partial u}{\partial x_n} \geq 0$ in $\Omega$. Since $\frac{\partial u}{\partial x_n}$ satisfies some linear equation, viz. $\Delta(\frac{\partial u}{\partial x_n}) + f'(u)\frac{\partial u}{\partial x_n} = 0$, one infers from the strong maximum principle that (see e.g. [PW]) $\frac{\partial u}{\partial x_n} > 0$ in $\Omega$.

To prove that $w^\lambda > 0$ in $\Omega_\lambda$, one uses the equation (1): $v^\lambda$ satisfies the same equation, $\Delta v^\lambda + f(v^\lambda) = 0$ in $\Omega_\lambda$. Hence, the function $w^\lambda$ satisfies some linear equation

(9) $$\Delta w^\lambda + c_\lambda(x) w^\lambda = 0 \quad \text{in} \quad \Omega_\lambda,$$

where

$$c_\lambda(x) = \frac{f(v^\lambda(x)) - f(u(x))}{v^\lambda(x) - u(x)}.$$

The function $c_\lambda$ is bounded in $L^\infty(\Omega_\lambda)$, uniformly with respect to $\lambda$ in bounded intervals. On the boundary, $w^\lambda$ satisfies

(10) $$w^\lambda \geq 0, \quad w^\lambda \not\equiv 0 \text{ on } \partial\Omega_\lambda.$$

In fact, $w^\lambda = 0$ on $T_\lambda$ whereas $w^\lambda > 0$ on the remaining part of the boundary, that is $\partial\Omega \backslash T_\lambda = \{x \in \partial\Omega, x_n < \lambda\}$.

We now use the method of [BN1]. It relies on the following version of the maximum principle in domains of small volume (see [BNV]).

Consider an elliptic operator

$$L = a_{ij}(x)\frac{\partial^2}{\partial x_i \partial x_j} + b_i(x)\frac{\partial}{\partial x_i} + c(x)$$

in a domain $\Omega$. Assume that $a_{ij} \in C(\overline{\Omega})$, $b_i$, $c \in L^\infty(\Omega)$, that the operator is uniformly elliptic, that is:

$$a_{ij}(x)\xi_i\xi_j \geq C_0|\xi|^2 \qquad \forall \xi \in \mathbb{R}^n, \ \forall x \in \Omega.$$

Let $b \geq \|b_i\|_{L^\infty(\Omega)}, \|c\|_{L^\infty(\Omega)}$. One says that $L$ satisfies the maximum principle in $\Omega$ if for any function $w$ which satisfies

$$\begin{aligned} Lw &\leq 0 \quad \text{in } \Omega \\ w &\geq 0 \quad \text{on } \partial\Omega \end{aligned}$$

one can infer that $w \geq 0$ in $\Omega$.

A sufficient condition for $L$ to satisfy the maximum principle is that $c(x) \leq 0$.

In general, when $c(x)$ is allowed to be positive, the maximum principle does not hold. However, if the domain is sufficiently small in measure, it still holds:

THEOREM 8. *There exists a constant $\delta > 0$ which depends on $n$, $c_0$, $b$ and $R > 0$ such that if the diameter of $\Omega$ is less than $R$ and*

$$meas(\Omega) < \delta$$

*then $L$ satisfies the maximum principle in $\Omega$.*

I refer the reader to [BNV] for a systematic study of conditions under which the maximum principle holds and some general results related to it.

*Step 1.* $w^\lambda > 0$ in $\Omega_\lambda$ for $0 < \lambda - \lambda_0$ small. Indeed, then, $\Omega_\lambda$ has small measure and since $\|c_\lambda\|_{L^\infty(\Omega_\lambda)}$ is bounded uniformly on compact sets of $\lambda$, the maximum principle applies (by Theorem 8 above). This could also be seen by using the fact that $\Omega_\lambda$ is *narrow* in the $x_n$ direction, if $0 < \lambda - \lambda_0$ is small (see [GNN1] or [BNV]). Therefore, $w_\lambda > 0$ in $\Omega_\lambda$ if $0 < \lambda - \lambda_0$ is small.

*Step 2.* Suppose $w^\lambda > 0$ in $\Omega_\lambda$ for all values of $\lambda$ in some interval $(\lambda_0, \mu)$. Then, by continuity, $w^\mu \geq 0$ in $\Omega_\mu$ and since $w^\mu \not\equiv 0$, we infer from (9) that $w^\mu > 0$ in $\Omega_\mu$. Let $K$ be a compact set included in $\Omega_\mu$ such that the measure of $\Omega_\mu \backslash K$ is small, namely,

$$\text{meas}(\Omega_\mu \backslash K) < \frac{\delta}{2}$$

for some small enough $\delta$. Then, since $\min_{x\in K} w^\mu(x) = \eta > 0$, if $\varepsilon > 0$ is sufficiently small, by continuity, we see that for all values of $\lambda \in [\mu, \mu + \varepsilon]$,

$$\min_{x\in K} w^\lambda(x) \geq \frac{\eta}{2} > 0.$$

Hence, $w^\lambda > 0$ on $K$ and, furthermore,

$$\text{meas}(\Omega_\lambda \backslash K) < \delta.$$

By the maximum principle for domains with small volume (see [BN1], [BNV]), if $\delta > 0$ is chosen sufficiently small, then the operator $\Delta + c_\lambda(x)$ satisfies the maximum principle in $\Omega_\lambda \backslash K$. Therefore, since $w^\lambda$ satisfies

$$\begin{aligned}
&\Delta w^\lambda + c_\lambda(x) w^\lambda = 0 \text{ in } \Omega_\lambda \backslash K \\
&w^\lambda \geq 0, w^\lambda \not\equiv 0 \text{ on } \partial\Omega_\lambda \\
&w^\lambda > 0 \text{ on } K
\end{aligned}$$

we see that $w^\lambda > 0$ in $\Omega_\lambda \backslash K$, hence in all of $\Omega_\lambda$.

We have thus shown that $w^\lambda > 0$ for all $\lambda$, not only in $(\lambda_0, \mu)$, but also in $(\lambda_0, \mu + \varepsilon]$.

We may now conclude. The above two steps show that $w^\lambda > 0$ for all $\lambda > \lambda_0$ and this implies that $\frac{\partial u}{\partial x_n} > 0$ in $\Omega$.

As I have said earlier, this proof does not extend to more general epigraph domains when coercivity condition (8) is not satisfied. There are two difficulties. Firstly, $\lambda_0$ is not necessarily finite and, second, $\Omega_\lambda$ is not bounded any more for all values of $\lambda$ and the problem becomes much more involved.

Next, let me observe that Theorem 6 implies, in the case of a half space, a *symmetry* and monotonicity result. Indeed, consider, as in the previous section, the case of a half space $\Omega = \{x \in \mathbb{R}^n; x_n > 0\}$ which corresponds to a flat graph $\varphi \equiv 0$. Then, for any direction $\xi \in \mathbb{R}^n \backslash \{0\}$ transversal to $\Omega$ and enterring $\Omega$, that is, such that $\xi_n > 0$, for any system of coordinates $y_1, \ldots, y_n$, there is a $\eta_n = \xi_n$ and one can read the half space as defined by $\Omega = \{y_n > \psi(x_1, \ldots, x_{n-1})$ with $\psi : \mathbb{R}^{n-1} \to \mathbb{R}$ Lipschitz.

Therefore, if $f$ satisfies conditions (4)-(7) and $\Omega$ is a half space, for any such direction $\xi$, Theorem 6 implies

$$\frac{\partial u}{\partial \xi} > 0 \text{ in } \Omega.$$

This implies symmetry. Indeed, let $x, x' \in \Omega$ be two points at equal distance of $\partial\Omega$, i.e. $x_n = x'_n$.

Let $e_n = (0, 0, ..., 1)$. Choosing appropriate directions $\xi$, the previous monotonicity property yields

$$u(x) < u(x' + \varepsilon\, e_n)$$

$$u(x') < u(x + \varepsilon\, e_n)$$

whence, by letting $\varepsilon \to 0$,

$$u(x) = u(x').$$

Actually, this result of symmetry in a half space under essentially conditions (4)-(7) had already been proved by S. Angenent [An] and by Clément and Sweers [CS](1) –with a slight improvement. They use different methods which cannot readily by extented to yield monotonicity in the more general case of a (non flat) Lipschitz graph as in Theorem 6.

S. Angenent and P. Clément and G. Sweers were led to this problem by still another motivation than the one described in this section. They were interested in describing the set of positive solutions for some singular perturbation problems

$$\text{(11)} \qquad \begin{cases} \varepsilon\, \Delta u + f(u) = 0, \quad u > 0, \text{ in } \Omega, \\ \qquad\qquad u = 0 \text{ on } \partial\Omega, \end{cases}$$

where $f$ satisfies (4)-(7). Angenent [An] shows that the positive solution of (11) is *unique* for small enough $\varepsilon > 0$.

Indeed, a rescaling of this equation near the boundary yields as a limit the problem (1). It is not difficult to see that under conditions (4)-(7), there is a unique solution of (1) such that $0 < u < \mu$. In fact symmetry shows that $u = u(x_n)$ satisfies the O.D.E.

$$\frac{d^2u}{dx_n^2} + f(u) = 0, \quad 0 < x_n < +\infty,$$

with conditions

$$u(0) = 0, \quad 0 < u_n(x) < \mu \text{ for } x_n > 0.$$

It is straightforward, under (4)-(7) to see that such a solution $u$ of the O.D.E. is uniquely determined.

Hence, this procedure shows that when $\varepsilon$ goes to 0, the boundary behaviour of solutions of (11) is in some sense uniquely determined. This information in turn is

(1) I owe to Guido Sweers to have brought these references to my attention.

crucial in the analysis of the solutions of (11) and to establish the uniqueness of the positive solution (see Angenent [An]).

## 4. Infinite cylinders

Another interesting case of infinite domain is that of cylinders. In this section, I assume that

$$\Omega = \mathbb{R} \times \omega$$

where $\omega \in \mathbb{R}^{n-1}$ is a bounded and smooth domain of $\mathbb{R}^{n-1}$. There are several types of symmetries one can be interested in. I will mention a few different results at the end of this section but here I will mainly focus on and start with *cylindrical symmetry*. That is, does the solution reflect the symmetries of the cross section domain $\omega$ ?

The answer to this problem is positive.

The solutions considered here may be unbounded, therefore the assumption made on $f$ is global:

(12) $f$ is (globally) Lipschitz continuous on $[0, +\infty)$.

Let me denote coordinates $x = (x_1, y)$ with $y = (y_1, \ldots, y_{n-1}) \in \omega \subset \mathbb{R}^{n-1}$.

The main result is the following

THEOREM 9. *Suppose $\omega$ is a domain in $\mathbb{R}^{n-1}$ which is convex in a direction, say $y_1$, of $\mathbb{R}^{n-1}$ and symmetric with respect to $y_1 = 0$. Then, any positive solution $u$ of (1) in the infinite cylinder $\Omega$ is symmetric with respect to $y_1 = 0$ and decreasing away from the symmetry hyperplane $y_1 = 0$. That is, $\frac{\partial u}{\partial y_1} > 0$ in $\Omega \cap \{y_1 > 0\}$. Consequently, if $\omega$ is a sphere in $\mathbb{R}^{n-1}$, then $u$ is cylindrically symmetric in the sense that $u(x, y) = u(x, |y|)$ and $\frac{\partial u}{\partial \rho}(x, y) < 0$ for $0 < \rho = |y|$ in $\Omega$.*

As I said, the solution may be bounded or not. Actually, we start by establishing a limitation on the growth of solutions of (1) in the infinite cylinder (under the assumption ((12) on $f$). This is the next result.

THEOREM 10. *Suppose $u$ is a positive solution of (1). Then $u$ has at most exponential growth. That is, there exist constants $C$, $\alpha > 0$ such that*

$$0 < u(x, y) \leq C\, e^{\alpha |x|} \text{ in } \Omega.$$

These two results are proved in the papers [BCN5] and [BCN6]. The second paper contains the case of a strip $n = 2$, that is $\Omega = \mathbb{R} \times (-1, 1)$. It turns out that this case is considerably more delicate and it will be explained why in the next section.

It is worth emphasizing that the positivity assumption in Theorem 10 is important. The result does not hold in general for solutions which change sign. Here is an example.

We consider $n = 2$, $\Omega = \mathbb{R} \times (-\pi, \pi)$ and the function

$$u(x, y) = \sum_{j=2k+1, k \in \mathbb{Z}} e^{j(x-j^2)} \cos jy.$$

The function $u$ is harmonic, changes sign and grows faster than exponential.

The method of proof of the above symmetry result (Theorem 9) in [BCN5] rests on the moving plane method. We use hyperplanes which are parallel to the axis of the cylinder, that is planes of the type $\{y_1 = \lambda\}$.

In addition to the usual arguments of the moving plane method (in the approach of [BN1]), we use two new tools which are of independent interest. Let me describe these two results now.

The first one deals with the maximum principle in infinite cylinders with a cross section of small measure. It is an adaptation of Theorem 8 to infinite cylinders. Consider an infinite cylinder –or a slab like domain– defined by

$$\Omega = \mathbb{R}^{n-j} \times \omega$$

where $\omega$ is a bounded and smooth domain of $\mathbb{R}^j$. We split coordinates of a point as $(x, y)$ with $x \in \mathbb{R}^j$ and $y \in \omega$.

In $\Omega$, suppose that the function $w$, with $w \in W^{2,n}_{\text{loc}}(\Omega) \cap C(\overline{\Omega})$, satisfies the inequalities

$$(12) \qquad \begin{cases} \Delta w + c(x, y)w \geq 0 \text{ in } \Omega \\ \qquad\qquad w \leq 0 \text{ on } \partial\Omega. \end{cases}$$

As before, the maximum principle applies when we can infer from (12) that, necessarily, $w \leq 0$ in $\Omega$. Here is a sufficient condition.

THEOREM 11. *Suppose that* $c \in L^\infty(\Omega)$ *with* $|c(x,y)| \le b \quad \forall (x,y) \in \Omega$, *and that*

$$|u(x,y)| \le Ce^{\gamma|x|} \text{ in } \Omega$$

*for some positive constants* $C$, $\mu$ *and* $b$. *There exists a constant* $\delta > 0$ *which only depends on* $n$, $j$, $b$ *and* $\mu$ *such that if*

$$|\omega| := \text{meas}(w) < \delta$$

*then, it follows from (12) that* $w \le 0$ *in* $\Omega$.

It turns out that this is a useful result in applying the moving plane approach to infinite cylinders.

The other result is linear. It concerns an extension of the Krylov-Safonov Harnack inequality up to the boundary. I state it now.

Consider a linear elliptic operator

$$L = \sum_{i,j=1}^{n} a_{ij}(x) \frac{\partial^2}{\partial x_i \partial x_j} + \sum_{i=1}^{n} b_i(x) \frac{\partial}{\partial x_i} + c(x)$$

where $x \in D$ and $D$ is a domain of $\mathbb{R}^n$, $a_{ij} \in C(\overline{D})$, $b_i$, $c \in L^\infty(D)$. The operator is assumed to be elliptic that is

$$\sum_{i,j=1}^{n} a_{ij}(x)\xi_i\xi_j \ge c_0|\xi|^2, \quad \forall \xi \in \mathbb{R}^n, \ \forall x \in \overline{D}.$$

Consider a function $u \in W^{2,p}_{\text{loc}}(D)$ for some $p > n$ satisfying

$$Lu = \chi f(x) \text{ in } D \tag{13}$$

where $\chi$ is a constant and $f \in L^\infty(D)$. Assume further that

$$\|b_i\|_{L^\infty}, \ \|c\|_{L^\infty}, \ \|f\|_{L^\infty} \le A.$$

The usual Krylov-Safonov inequality deals with positive solution of (13) and asserts that for a compact subset $K$ of $D$, when, say, $\chi = 0$ there exits a constant $C$ such that

$$\sup_{x \in K} u(x) \le C \inf_{x \in K} u(x).$$

The extension which we present in [BCN5] is to have such an inequality up to the boundary, or portion of it, for functions which vanish there. Such an inequality turns out to be very useful. For greater generality, this inequality is stated even when $\chi$ is not 0 in (13).

THEOREM 12. *Let $D$ be a domain of $\mathbb{R}^n$ and $\Sigma$ be an open set of $\partial D$ which is assumed to be smooth. Suppose $u$ is a positive function in $\Omega$, $u \in W^{2}_{\mathrm{loc}}(D \cup \Sigma)$ for some $p > n$, such that $u$ vanishes on $\Sigma$ and $u$ satisfies (13). Let $K$ be a compact subset of $D$ and $H$ a compact subset of $D \cup \Sigma$. Then,*

$$\sup_{x \in H} u(x) \le C(\inf_{x \in K} u(x) + \chi)$$

*where $C$ is a constant which only depends on $D$, $\Sigma$, $K$, $H$, $C_0$ and $A$.*

As I indicated above, the solutions which we consider here may be unbounded. Therefore, the functions $f$ are restricted to satisfy the global Lipschitz condition (13). It is not known whether this condition can be dropped.

*Open problem 5.* Does the result of Theorem 9 above remain true if one assumes only that $f$ is locally Lipschitz continuous?

Of interest is also the question to know whether there are growth restrictions for positive solutions of equation (1) in infinite cylinders $\Omega$ which extend Theorem 10. One may consider this question with an assumption on the growth of $f$ e.g. $f(u) \simeq Cu^p$ as $u \to \infty$ for some $C > 0$ and $p \ge 1$.

There have been several recent other works concerned with qualitative properties of semi-linear elliptic equations in infinite cylindrical domains. In my paper with L. Nirenberg [BN2], we investigate monotonicity properties for travelling fronts, that is solutions $u$ of equations of the type

$$\begin{cases} \Delta u - \beta(y)\dfrac{\partial u}{\partial x} + f(y,u) = 0 \quad \text{in } \Omega = \mathbb{R} \times \omega, \\ \dfrac{\partial u}{\partial \nu} = 0 \quad \text{on } \partial\Omega, \\ u(-\infty, y) = 0, \quad u(+\infty, y) = 1. \end{cases}$$

We prove that if $0 < u < 1$, then $u$ is monotonic in the $x_1$ direction, i.e. $\frac{\partial u}{\partial x_1} > 0$ in $\Omega$. We also consider other types of boundary conditions like Dirichlet conditions.

Yet another result is that of *symmetry* of travelling waves (or "pulses") which are positive solutions of (1) in $\Omega = \mathbb{R} \times \omega$ which satisfy

$$u(x, y) = 0 \text{ as } x \to \pm\infty.$$

We prove in [BN3] that, under some conditons on $f$ and $\beta$, the solution is symmetric in $x$ up to a shift of the origin along the axis. That is, there exists $x_0 \in \mathbb{R}$ such that

$$v(x, y) = u(x + x_0, y)$$

is symmetric with respect to $x$ and satisfies $\frac{\partial v}{\partial x} < 0$ for $x > 0$ in $\Omega$.

## 5. Infinite slab type domains

One can also obtain symmetry results for more general cylindrical domains

$$\Omega = \mathbb{R}^{n-j} \times \omega$$

which one can think of as infinite slabs or strip like domains. I already mentioned such domains in Theorem 11 above. Here, $\omega$ is a bounded and smooth domain of $\mathbb{R}^j$. The coordinates of a point $X \in \Omega$ are split, as before, as $X = (x, y)$ with $x = (x_1, ..., x_{n-j}) \in \mathbb{R}^{n-j}$, $y = (y_1, ..., y_j) \in \omega$. Let us now consider problem 1 in this infinite slab $\Omega$. It is also assumed here that $f$ is globally Lipschitz continuous (condition (12)).

Firstly, the restriction on the growth of a positive solution of (1) in $\Omega$ is unchanged with respect to the previous case when $n - j = 1$. That is, Theorem 9 still holds whatever $j$ is, $1 \leq j \leq n - 1$, which means that $u$ has at most exponential growth.

We also prove a symmetry result in [BCN5], [BCN6]. However, when $\omega$ is an interval, that is, $j = 1$, there is a restriction.

Theorem 13. *Suppose $\omega$ is convex in the $y_1$ direction and that it is symmetric about the hyperplane $\{y_1 = 0\}$. Assume that either $j \geq 2$ or that $n = 2$ or, if $n \geq 3$ and $j = 1$, that $f(0) \geq 0$. Then, $u$ is symmetric in $y_1$ and $\frac{\partial u}{\partial y_1} < 0$ for $y_1 > 0$ in $\Omega$.*

Notice that the case $n = 2$, $j = 1$, corresponding to an infinite strip in the plane is contained in Theorem 9.

It turns out that the case $j = 1$, when $\omega$ is an interval, and $f(0) < 0$, is considerably more delicate. This is why we have obtained this symmetry result in that case so far only when $n = 2$.

Essentially, the difficulty rests on the existence of solutions $b(y)$ of the problem

$$\begin{cases} b''_{yy} + f(b) = 0, \quad b \geq 0 \quad \text{in} \quad \omega \times (-L, L) \\ b(-L) = b(L) = 0, \end{cases} \tag{15}$$

with $b$ actually vanishing at some points in the interval $(-L, L)$. Such solutions do not exist when $f(0) \geq 0$ whereas it is easy to construct them when $f(0) < 0$. This phenomenon does not happen in dimension higher than one.

In other words, consider a solution $b$ of

$$\begin{cases} \Delta b + f(b) = 0 & \text{in}\, \omega \subset \mathbb{R}^j, \\ b = 0 & \text{on}\, \partial\omega, \end{cases}$$

with $b \geq 0$, $b \not\equiv 0$. Assume that $\omega$ is a bounded and smooth domain in $\mathbb{R}^j$ and that $\omega$ is convex in some direction, say $y_1$, and symmetric about the hyperplane $\{y_1 = 0\}$. Then, if the dimension $j$ is $\geq 2$, then $b > 0$ in $\omega$. This fact was observed by Castro and Shivaji who stated it when $\omega$ is a sphere. One can extend their proof to get this more general statement.

*Open problem 6.* Does Theorem 13 also hold when $j = 1$, $n \geq 3$ and $f(0) < 0$ ?

In fact, in [BCN6], for the strip, since there are solutions $b(y)$ of (15), in some sense one has to show in particular that there is no solution $u$ of (1) in the strip $\mathbb{R} \times (-L, L)$ which is positive and such that $u(x, y) \to b(y)$ as $x$ approaches $+\infty$ or $-\infty$. To exclude such solutions turns out to be delicate.

## 6. Results of the Serrin type in unbounded domains

The method of moving planes was developed, following Alexandrov [A] in a celebrated paper of Serrin [S]. In this paper, Serrin considers an equation associated with overdetermined boundary conditions:

$$\begin{cases} \Delta u + f(u) = 0 \quad \text{in}\, \Omega, \quad u > 0 \quad \text{in}\, \Omega, \\ u = 0 \quad \text{and} \quad \dfrac{\partial u}{\partial \nu} = \alpha \quad \text{on}\, \partial\Omega, \end{cases} \tag{16}$$

where $\Omega$ is supposed to be a smooth and bounded domain of $\mathbb{R}^n$ with unit outward normal $\nu$ and $\alpha$ is a constant. We also assume that $f : \mathbb{R}_+ \to \mathbb{R}$ is locally Lipschitz continuous.

Serrin's result asserts the *symmetry of the domain* $\Omega$ –in fact its uniqueness– for which there exists a solution of such a problem.

THEOREM (SERRIN). *Let $\Omega$ be a smooth and bounded domain. If there exists a positive solution $u$ of the overdetermined boundary value problem (16) with $\alpha$ being constant, then, $\Omega$ is a sphere.*

This Theorem has an interpretation in terms of the electrostatic potential (see [S]) to which I will return in a moment. It also has various extensions, in particular to some quasilinear equations, when, in (16), $f = f(u, \nabla u)$.

Only very recently have there been some versions of this result in the framework of unbounded domains. These results deal mostly with *exterior* domains.

Let me describe this problem now. Consider a bounded domain $\Omega$ of $\mathbb{R}^n$ such that $\mathbb{R}^n \backslash \overline{\Omega}$ is also a domain, that is, the set $\mathbb{R}^n \backslash \overline{\Omega}$ is connected. In this domain, $\mathbb{R}^n \backslash \overline{\Omega}$, which is referred to as an exterior domain, consider an overdetermined boundary value problem

$$(17) \qquad \begin{cases} \Delta u + f(u) = 0 & \text{in } \mathbb{R}^n \backslash \overline{\Omega} \\ u = a, \quad \dfrac{\partial u}{\partial \nu} = c & \text{on } \partial\Omega \\ \underline{u} \leq u < u_{|\partial\Omega} = a & \text{in } \mathbb{R}^n \backslash \overline{\Omega}. \end{cases}$$

Here, $\nu$ is the inward (to $\Omega$) unit normal on $\partial\Omega$, $a$ and $c$ are constants.

The goal here is to establish that when such a solution exists, then, $\Omega$ is a sphere. A particular case of this problem –when $f \equiv 0$– has a natural interpretation in terms of the electrostatic potential.

Consider a conducting body $\overline{\Omega}$, in which the electrostatic potential $V$ is constant, $V \equiv a$ and assume that there is a uniform electric charge on the surface $\partial\Omega$, that is, $\frac{\partial V}{\partial \overrightarrow{\nu}} = -\overrightarrow{E} \cdot \overrightarrow{\nu}$ is constant. In the exterior of $\overline{\Omega}$, $V$ satisfies Laplace's equation $\Delta V = 0$ and $V$ decays to 0 at infinity. Then, the assertion that $\Omega$ is a sphere is a formal justification of the classical effect according to which, electric charges accumulate in pointed parts of the conductor and have a tendency to rarefy in troughs.

The first result of this kind is due to W. Reichel [R]. He proves the following

THEOREM (REICHEL). *Suppose $\Omega$ is a $C^2$ domain and there exists a solution $u$ of (17) in the exterior domain $\mathbb{R}^n \backslash \overline{\Omega}$. Assume furthermore that $u(x) \to 0$ uniformly as $|x| \to \infty$ and that $f(s)$ is decreasing in some right neighborhood $s \in [0, \delta]$ of 0. Then, $\Omega$ is a sphere and $u$ is spherically symmetric and decreasing away from the sphere, i.e. $\frac{\partial u}{\partial r} < 0$ in $\mathbb{R}^n \backslash \overline{\Omega}$.*

There is another result by Reichel concerning quasilinear equations

$$\Delta u + f(u, \nabla u) = 0.$$

(see [R]).

Very recently, A. Aftalion and J. Busca [AB] have obtained another result under different assumptions on $f$. They assume that $f$ satisfies

$$f(t) \geq 0 \quad \text{for} \quad t \geq 0. \tag{18}$$

$$t^{\frac{(n+2}{n-2}}) f(t) \text{ is a nonincreasing function of } \quad t \geq 0. \tag{19}$$

THEOREM (AFTALION - BUSCA). *If $f$ satisfies (18)-(19) and there exists a solution of (17) –it suffices to assume $u \leq a$ instead of $u < a$– Then $\Omega$ is a sphere and $u$ is spherically symmetric and decreasing away from $\Omega$.*

This covers for instance the case of

$$f(u) = u^p \quad \text{with } 1 \leq p \leq \frac{n+2}{n-2}$$

or

$$f(u) = \sum_{i=1}^{k} \alpha_i u^{p_i}$$

with $\alpha_i > 0$ and $1 \leq p_i \leq \frac{n+2}{n-2}$.

The method by Reichel is that of moving planes. Essentially, the assumptions that $f'(0) < 0$ and that $u \to 0$ uniformly as $|x| \to \infty$, allow one to apply the maximum principle at infinity. For instance, if $u$ and $v$ are solutions, then the difference satisfies some equation

$$\Delta(u-v)+c(x)(u-v)=0$$

with $c(x) < 0$ if $|x|$ is sufficiently large (hence, for instance, if $x_1 > \lambda$ with $\lambda$ sufficiently large).

Exploiting this fact, Reichel was able to implement the moving planes method.

For Aftalion and Busca however, who consider the case $f'(0) \geq 0$, this does not apply. Dealing with the infinite region of the domain is a more delicate. They first perform Kelvin transform and then use the reflection in moving planes for this transform. This is why they require assumptions (18) and (19). They also use a nice device of obtaining the reflection in a hyperplane as a limit of an inversion with respect to a sphere such that one point of the boundary –where the sphere is tangent to the given hyperplane– is kept fixed, while the center of the sphere is sent to infinity. Lastly, they also use the maximum principle in domains with small volume.

Aftalion and Busca also obtain a radial result when assuming a priori that $\Omega$ is a sphere, say $B_R(0)$. Here is their result.

THEOREM (AFTALION AND BUSCA [AB]). *Under the assumption (18)-(19) suppose $u > 0$ is a solution of $\Delta u + f(u) = 0$ in $\mathbb{R}^n \backslash \overline{B_R(0)}$ with $u = a$ (constant) on $\partial B_R(0)$ and $0 \leq u < a$. Then, $u$ is spherically symmetric and $\frac{\partial u}{\partial r} < 0$ in $\mathbb{R}^n \backslash \overline{B_R(0)}$.*

It is natural to ask whether such results hold in greater generality.

*Open problem 7.* Suppose $u$ is a bounded solution of $\Delta u + f(u) = 0$ in a smooth exterior domain $\mathbb{R}^n \backslash \overline{\Omega}$ with $u = a$ and $\frac{\partial u}{\partial \nu} = c$ on $\partial\Omega$ and $u < a$ outside $\overline{\Omega}$. Does this imply in general that $\Omega$ is a sphere?

Next, let me turn to the case when $\Omega$ is *unbounded.* I am still considering problem (17). What can be said about such a region $\Omega$ ? This is a new kind of problem. Here is a very recent and first result in this direction.

Suppose $\Omega$ is bounded by a smooth globally Lipschitz graph $\varphi$:

$$\Omega = \{x \in \mathbb{R}^n; x_n > \varphi(x_1, ..., x_{n-1})\}$$

with $\varphi \in C^2(\mathbb{R}^{n-1}, \mathbb{R})$ and let me assume for simplicity that $\varphi$ is globally bounded in $C^2$ norm.

THEOREM 14. *Assume conditions (4)-(7) on $f$ with $\mu = a$. Suppose that in the exterior of this domain $\Omega$ which bounds the graph of $\varphi$ there is a solution of*

$$\begin{cases} \Delta u + f(u) = 0, & 0 < u < a, \quad \text{in } \mathbb{R}^n \backslash \overline{\Omega} \\ u = a, & \dfrac{\partial u}{\partial \nu} = c \quad \text{on } \partial\Omega, \end{cases}$$

*with $c$ being a constant. Then, $\Omega$ is a half space, that is $\varphi$ is constant.*

This result is obtained by using the sliding method combined with some of the arguments in [BCN4].

I beleive that Theorem 14 is the first in a series of results of this type for various classes of domains $\Omega$ and functions $f$.

To conclude this talk, and in view of all these results, it is very tempting to formulate the following general.

*Conjecture.* Let $\Omega$ be a smooth domain with inward unit normal $\nu$. Assume that the region $\mathbb{R}^n \backslash \overline{\Omega}$ is also connected and that there is a solution of

$$\Delta u + f(u) = 0, \quad 0 < u \leq a \quad \text{in } \mathbb{R}^n \backslash \overline{\Omega}$$

with

$$u = a, \quad \frac{\partial u}{\partial \nu} = c \quad \text{on } \partial\Omega$$

($a$ and $c$ are constants). Then, either $\Omega$ is bounded and it is a sphere or it is unbounded and it is a half space.

## REFERENCES


[A] A.D. Alexandrov, A characteristic property of the spheres, *Ann. Mat. Pura Appl.*, 58 (1962), p. 303–354.

[AB] A. Aftalion and J. Busca, Radial symmetry of overdetermined boundary value problems in exterior domains. *To appear.*

[An] S.B. Angenent, Uniqueness of the solution of a semilinear boundary value problem. *Math. Ann.*, 272 (1985), p. 129–138.

[BN1] H. Berestycki and L. Nirenberg, On the method of moving planes and the sliding method. *Boletim Soc. Brasil. de Mat. Nova Ser.*, 22, (1991), p. 1–37.

[BN2] H. Berestycki and L. Nirenberg, Travelling wave in cylinders. *Ann. Inst. H. Poincaré, Anal. non Linéaire,* 9, (1992), p. 497–572.

[BN3] H. Berestycki and L. Nirenberg, Some qualitative properties of solutions of semilinear elliptic equations in a cylinder, *in Analysis Et Cetera,* P. Rabinowitz et al ed., Academic press, New York (1990), p. 115–164.

[BCN1] H. Berestycki, L. Caffarelli and L. Nirenberg, Uniform estimates for regularization of free boundary problems, *in Analysis and Partial Differentail Equations,* C. Sadovsky ed., M. Dekker, New York (1990), Lecture Notes in Pure and Applied Math. vol. 122, p. 567–619.

[BCN2] H. Berestycki, L. Caffarelli and L. Nirenberg, Symmetry for elliptic equations in a half space. *in Boundary value problems for partial differential equations and applications,* volume dedicated to E. Magenes. J.L. Lions et al. ed., Masson, Paris (1993), p. 27–42.

[BCN3] H. Berestycki, L. Caffarelli and L. Nirenberg, *In preparation.*

[BCN4] H. Berestycki, L. Caffarelli and L. Nirenberg, Monotonicity for elliptic equations in an unbounded Lipschitz domain. *In preparation.*

[BCN5] H. Berestycki, L. Caffarelli et L. Nirenberg, Inequalities for second order elliptic equations with applications to unbounded domains I. *Duke Math. J., to appear.*

[BCN6] H. Berestycki, L. Caffarelli et L. Nirenberg Inequalities for second order elliptic equations with applications to unbounded domains. II: Symmetry in infinite strips. *In preparation.*

[BNV] H. Berestycki, L. Nirenberg and S. R. S. Varadhan, The principal eigenvalue and Maximum principle for second order elliptic operators in general domains. *Comm. Pure Applied Math.,* 47, (1994), p. 47–92.

[CS] Ph. Clément and G. Sweers, Existence and multiplicity results for a semilinear elliptic eigenvalue problem, p. 97 –121.

[D] E.N. Dancer, Some notes on the method of moving planes, *Bull. Australian Math. Soc.,* 46 (1992), p. 425–434.

[EL] M. Esteban and P.L. Lions, Existence and non existence results for semilinear elliptic problems in unbounded domains, *Proc. Royal Soc. Edinburgh,* 93A

(1982), p. 1–14.

[GNN1] B. Gidas, W. M. Ni, and L. Nirenberg, Symmetry and related properties via the maximum principle. *Comm. Math. Phys.,* 6, 1981, p. 883–901.

[GNN2] B. Gidas, W. M. Ni, and L. Nirenberg, Symmetry of positive solutions of nonlinear elliptic equations in $\mathbb{R}^n$. *Math. Anal. Appl. Part A, Advances in Math. Suppl. Studies,* 7A, 1981, p. 369–402.

[L] Congming Li, Monotonicity and symmetry of solutions of fully nonlinear elliptic equations on unbounded domains. *Commun. Part. Diff. Equ.,* 16 (1991), p. 585–615.

[MM] L. Modica and S. Mortola, Some entire solutions in the plane of nonlinear Poisson equations, *Boll. U.M.I.,* 17B (1980), p. 614–622.

[PW] M. H. Protter and H. F. Weinberger, *Maximum Principles in Differential Equations.* Prentice-Hall, Englewood Cliffs, New Jersey, 1967.

[R] W. Reichel, Radial symmetry by moving planes for semilinear elliptic boundary value problems on exterior domains. *To appear.*

[S] J. Serrin, A symmetry theorem in potential theory. *Arch. Rat. Mech. Anal.,* 43, 1971, p.304–318.

H. Berestycki
Université Paris VI
Lab. Analyse Numérique
4 Pl. Jussieu
75252 Paris Cedex 05
France

L. BOCCARDO

# Some nonlinear Dirichlet problems in $L^1$ involving lower order terms in divergence form

## 1. INTRODUCTION AND HYPOTHESES

This note deals with existence and qualitative properties of solutions of the following boundary value problem

$$\begin{cases} A(u) = f - div\, \Phi(u) & \text{in } \Omega, \\ u = 0 & \text{on } \partial\Omega, \end{cases} \tag{1.1}$$

or, more in general, of the differential problem

$$\begin{cases} A(u) = f - div[F + \Phi(u)] & \text{in } \Omega, \\ u = 0 & \text{on } \partial\Omega, \end{cases} \tag{1.2}$$

where $A$ is a Leray-Lions operator defined from $W_0^{1,p}(\Omega)$ into its dual, $\Omega$ is a bounded open set of $\mathbf{R}^N$, and $1 < p < N$. The partial differential operator $A$ is defined by

$$A(v) = -div\,(a(x, v, \nabla v))$$

where $a : \Omega \times \mathbf{R} \times \mathbf{R}^N \to \mathbf{R}^N$ is a Caratheodory function. We assume that there exist two real positive constants $\alpha, \beta$ and a function $d(x) \in L^{p'}(\Omega)$ such that for every $s \in \mathbf{R}$, $\xi \in \mathbf{R}^N$, $\eta \in \mathbf{R}^N$ $(\xi \neq \eta)$, and for almost every $x \in \Omega$,

$$a(x, s, \xi)\xi \geq \alpha|\xi|^p, \tag{1.3}$$

$$|a(x, s, \xi)| \leq \beta\big(d(x) + |s|^{p-1} + |\xi|^{p-1}\big), \tag{1.4}$$

$$[a(x, s, \xi) - a(x, s, \eta)][\xi - \eta] > 0. \tag{1.5}$$

Concerning the right hand side of (1.1), (1.2) we assume that

$$f \in L^1(\Omega), \tag{1.6}$$

$$F \in (L^{p'}(\Omega))^N, \tag{1.7}$$

$$\Phi \in C^0(\mathbf{R}, \mathbf{R}^N). \tag{1.8}$$

We point out that no growth hypothesis is assumed on the function $\Phi$. This implies that for a function $v \in W_0^{1,r}(\Omega)$, $r > 1$, the term $div\, \Phi(v)$ may be meaningless, even as a distribution.

If $\Phi = 0$, $F = 0$, the existence of a weak solution in the Sobolev space $W_0^{1,q}(\Omega), q < \tilde{p} = \frac{(p-1)N}{N-1}$ has been studied (also if the right hand side is a Radon measure) by G. Stampacchia ([St]), if $A$ is linear and in some papers in collaboration with T. Gallouet, if $A$ is nonlinear ([BG1], [BG2], [BBGV], [BGV]; see also [Dv]). Of course, if $\Phi = 0$, $u$ is a weak solution of (1.1) if

$$\begin{cases} u \in W_0^{1,q}(\Omega), q < \tilde{p} = \frac{(p-1)N}{N-1} : \\ \displaystyle\int_\Omega a(x,u,\nabla u)\nabla\varphi = \int_\Omega f\varphi \qquad\qquad \forall\varphi \in \mathcal{D}(\Omega). \end{cases} \tag{1.9}$$

Even in the linear case, the solution is not necessarily unique: a counterexample can be found in [Se] (for a complete description see [P]). However, if $A$ is monotone (roughly speaking: if $a(x,s,\xi)$ is independent of $s$, see [BD]) the uniqueness of the solution is proved by adding to the weak formulation a so called entropy condition.

**Definition 1.1.** ([BBGV]) If $\Phi = 0$ and $p > 2 - \frac{1}{N}$, an entropy solution of (1.1) is a function $u$ such that

$$\begin{cases} u \in W_0^{1,q}(\Omega), q < \tilde{p} = \frac{(p-1)N}{N-1}, \\ T_k(u) \in W_0^{1,p}(\Omega), \forall k > 0 : \\ \displaystyle\int_\Omega a(x,\nabla u)\nabla T_k[u-\varphi] \le \int_\Omega f T_k[u-\varphi] \\ \qquad \forall\varphi \in W_0^{1,p}(\Omega) \cap L^\infty(\Omega), \end{cases} \tag{1.10}$$

where

$$T_k(t) = \begin{cases} t, & |t| \le k, \\ k\frac{t}{|t|}. & |t| > k. \end{cases}$$

Thus, while it is not possible to use the solution $u$ as test function in (1.9), because $u$ does not belong to $W_0^{1,p}(\Omega)$, an important role is played by $T_k(u)$ (see

[BG 5] for the first utilization of these functions in the weak formulations of some elliptic problems) and by the test functions

$$T_k[u-\varphi], \qquad \varphi \in W_0^{1,p}(\Omega) \cap L^\infty(\Omega),$$

because both belong to $W_0^{1,p}(\Omega)$. Moreover not every $v \in W_0^{1,p}(\Omega)$ can be used as test function in (1.9) since the term $a(x,u,\nabla u)\nabla v$ may not belong to $L^1(\Omega)$. Another boundary value problem where the use of nonlinear bounded test functions depending on $u$ is needed in order to give a good definition of solution is the Dirichlet problem (1.1), studied in [BDGM] if $f \in W^{-1,p'}(\Omega)$, by means of the so called renormalized solutions.

The aim of this note is to show the existence of solutions for problem (1.1), if $f \in L^1(\Omega)$ and $\Phi \neq 0$, in the framework of entropy solutions.

The main tools of this approach are an *a priori* estimate in the Sobolev space $W_0^{1,q}(\Omega)$ (see also [BG1], [BG2]) and the almost everywhere convergence of the gradients of approximate equations (see also [BG1], [BG2], [Dv], [BM]).

Another approach can be found in two recent papers by F. Murat ([M], if $A$ is linear, and [LM]) in the framework of renormalized solutions of [BDGM].

## 2. EXISTENCE

**Theorem 2.1.** *Let $p$ be such that $2-\frac{1}{N} < p < N$. Let $A$ be as before, let $f$ be a summable function on $\Omega$ and define $\Phi_n(s) = \Phi[T_n(s)]$. Let $\{f_n\}$ be any sequence of smooth functions that converges to $f$ in $L^1(\Omega)$ and let $u_n \in W_0^{1,p}(\Omega)$ be, for every $n$ in $\mathbf{N}$, a weak solution of the boundary value problem*

$$\begin{cases} A(u_n) = f_n - div\, \Phi(u_n) & \text{in } \Omega, \\ u_n = 0 & \text{on } \partial\Omega. \end{cases} \tag{2.1}$$

*Then for every $q < \tilde{p} = \frac{N(p-1)}{N-1}$ there exists a positive constant $c_q$ (independent of $n$) such that*

$$\|u_n\|_{W_0^{1,q}(\Omega)} \leq c_q \qquad \forall n \in \mathbf{N}. \tag{2.2}$$

*Moreover, $u_n$ converges in $W_0^{1,q}(\Omega)$ to a function $u$ that is an entropy solution of the Dirichlet problem (1.1), that is*

$$\begin{cases} u \in W_0^{1,q}(\Omega), q < \tilde{p} = \frac{(p-1)N}{N-1}, \\ T_k(u) \in W_0^{1,p}(\Omega), \forall k > 0 : \\ \displaystyle\int_\Omega a(x,u,\nabla u)\nabla T_k[u-\varphi] \leq \int_\Omega f T_k[u-\varphi] + \int_\Omega \Phi(u)\nabla T_k[u-\varphi] \\ \qquad \forall \varphi \in W_0^{1,p}(\Omega) \cap L^\infty(\Omega). \end{cases} \tag{2.3}$$

**Remark 2.2.** Observe that if $p > N$, then $L^1(\Omega) \subset W^{-1,p'}(\Omega)$. Thus the above existence result is the result of [BGi], since the solution is bounded.

**Proof.**

To prove the existence theorem we proceed by steps.

**Step 1 - Approximate solutions**

*(Allegro vivace assai)*

From the Leray-Lions existence theorem ([LL]) there exists at least one solution $u_n$ of the approximate Dirichlet problem (2.1). The choice $f_n = T_n(f)$ is an example of an approximating sequence $f_n$.

**Step 2 - A priori estimates**

*(Minuetto)*

Now we prove that the sequence $u_n$ is bounded in $W_0^{1,q}(\Omega)$ for any $q < \tilde{p}$. Let $\psi_k(t)$ be the real valued function defined by

$$\psi_k(t) = \begin{cases} 0 & \text{if } t \in [0,k], \\ t-k & \text{if } t \in (k, k+1], \\ 1 & \text{if } t \in (k+1, +\infty), \\ -\psi_k(-t) & \text{otherwise.} \end{cases} \tag{2.4}$$

By the Divergence Theorem (see also [BDGM]), we get

$$\int_\Omega \Phi_n(u_n)\ \nabla\psi_k(u_n) \ = 0.$$

Thus, the use of $\psi_k(u_n)$ in (2.1) implies that

$$\alpha \int_{B_k} |\nabla u_n|^p \leq \int_\Omega a(x,u_n,\nabla u_n)\nabla\psi_k(u_n) = \int_\Omega f_n \psi_k(u_n) \leq \|f\|_{L^1(\Omega)}. \tag{2.5}$$

where

$$B_k = \{x \in \Omega : k \le |u_n(x)| < k+1\}. \tag{2.6}$$

Thus, for every $\lambda > 1$, we have

$$\int_\Omega \frac{|\nabla u_n|^p}{(1+|u_n|)^\lambda} = \sum_{k=0}^{\infty} \int_{B_k} \frac{|\nabla u_n|^p}{(1+|u_n|)^\lambda} \le \sum_{k=0}^{\infty} \frac{\|f\|_{L^1(\Omega)}}{\alpha(1+k)^\lambda} = c_1. \tag{2.7}$$

Let $q < p$; by Sobolev embedding and Hölder inequality,

$$c_0 \left( \int_\Omega |u_n|^{q*} \right)^{\frac{q}{q*}} \le \int_\Omega |\nabla u_n|^q = \int_\Omega \frac{|\nabla u_n|^q}{(1+|u_n|)^{\frac{\lambda q}{p}}} (1+|u_n|)^{\frac{\lambda q}{p}}$$

$$\le \left( \int_\Omega \frac{|\nabla u_n|^p}{(1+|u_n|)^\lambda} \right)^{\frac{q}{p}} \left( \int_\Omega (1+|u_n|)^{\frac{\lambda q}{p-q}} \right)^{1-\frac{q}{p}} \le (c_1)^{\frac{q}{p}} c_2 \left( 1 + \int_\Omega |u_n|^{\frac{\lambda q}{p-q}} \right)^{1-\frac{q}{p}}.$$

The previous inequalities give an *a priori* estimate of the sequence $u_n$ in the <u>large</u> Sobolev space $W_0^{1,q}(\Omega), q < \tilde{p}$, if:

(i) $\frac{q}{q^*} > 1 - \frac{q}{p}$, that is, if $p < N$;

(ii) $q^* = \frac{\lambda q}{p-q}$ and $\lambda > 1$, that is, if $q < \tilde{p}$.

Hence, the above *a priori* estimate implies that there exists a function $u \in W_0^{1,q}(\Omega)$ and a subsequence, still deboted by $u_n$, such that (for every $q$ such that $1 \le q < \tilde{p}$)

$$u_n \to u \qquad \text{weakly in } W_0^{1,q}(\Omega),\ \forall q < \tilde{p}, \quad \text{strongly in } L^r(\Omega),\ \forall r < \frac{(p-1)N}{N-p}.$$

Moreover, the use of $T_k(u_n)$ as test function in (2.1) implies that the sequence $T_k(u_n)$ is bounded in the Sobolev space $W_0^{1,p}(\Omega)$ and so

$$T_k(u_n) \to T_k(u) \qquad \text{weakly in } W_0^{1,p}(\Omega),\ \forall k > 0. \tag{2.8}$$

**Step 3 - $L^1$ compactness of the sequence $\nabla u_n$**

*(Andante cantabile)*

Fix $q$ such that $1 < q < \tilde{p}$. Let $0 < \theta < \frac{q}{p}$ $(< 1)$ and $k > 0$. Consider

$$I_{\Omega,n} = \int_\Omega \{[a(x,u_n,\nabla u_n) - a(x,u_n,\nabla u)]\nabla(u_n - u)\}^\theta \tag{2.9}$$

We shall prove that the previous integral converges to zero. Indeed, it is equal to

$$\int_{C_k} \{[a(x,u_n,\nabla u_n) - a(x,u_n,\nabla u)]\nabla(u_n - u)\}^\theta$$

$$+ \int_{A_k} \{[a(x,u_n,\nabla u_n) - a(x,u_n,\nabla u)]\nabla(u_n - u)\}^\theta$$

$$= I_{C_k,n} + I_{A_k,n}, \tag{2.10}$$

where

$$C_k = \{x \in \Omega : |u(x)| \le k\}, \qquad A_k = \{x \in \Omega : |u(x)| > k\}.$$

We can write $I_{C_k,n}$ as

$$\int_{C_k} \{[a(x,u_n,\nabla u_n) - a(x,u_n,\nabla T_k(u))]\nabla(u_n - T_k(u))\}^\theta,$$

which is smaller than

$$\int_\Omega \{[a(x,u_n,\nabla u_n) - a(x,u_n,\nabla T_k(u))]\nabla(u_n - T_k(u))\}^\theta = J_{\Omega,n},$$

since the integrand is positive. Let $r = \frac{q}{p\,\theta}$. The change of the integration set in the first integral, the use of Hölder inequality (with exponents $r$ and $r'$) and (1.4) in the second one imply that

$$I_{C_k,n} + I_{A_k,n} \le J_{\Omega,n} + c_1 \left[\int_{A_k} (|\nabla u_n| + |\nabla u|)^{pr\theta}\right]^{\frac{1}{r}} meas(A_k)^{1-\frac{1}{r}}.$$

By means of the *a priori* estimate, proved in the previous Step, we get

$$I_{C_k,n} + I_{A_k,n} \le J_{\Omega,n} + c_2\, meas(A_k)^{1-\frac{1}{r}} = J_{\Omega,n} + \omega_1(k).$$

Where we have denoted by by $\omega_i(k)$ some quantities such that $\lim_{k\to\infty}\omega_i(k) = 0$. Now we study the behaviour of $J_{\Omega,n}$ in (2.11); it can be splitted as ($j \in \mathbf{N}$)

$$\int_{\{|u_n(x) - T_k(u)| \le j\}} \{[a(x,u_n,\nabla u_n) - a(x,u_n,\nabla T_k(u))]\nabla[u_n - T_k(u)]\}^\theta$$

$$+\int_{\{|u_n(x)-T_k(u)|>j\}}\{[a(x,u_n,\nabla u_n)-a(x,u_n,\nabla T_k(u))]\nabla[u_n-T_k(u)]\}^{\theta}. \quad (2.11)$$

The first integral can be written as

$$\int_{\Omega}\{[a(x,u_n,\nabla u_n)-a(x,u_n,\nabla T_k(u))]\nabla T_j[u_n-T_k(u)]\}^{\theta}.$$

Then we use the Hölder inequality (with exponents $\theta$ and $\frac{1}{1-\theta}$) and the *a priori* estimate of the previous Step in order to estimate (2.11) as

$$\left(\int_{\Omega}[a(x,u_n,\nabla u_n)-a(x,u_n,\nabla T_k(u))]\nabla T_j[u_n-T_k(u)]\quad\right)^{\theta}\quad(meas\ \Omega)^{1-\theta}$$

$$+c_3\ meas\{x\in\Omega:|u_n(x)-T_k(u(x))|>j\}^{1-\frac{1}{r}}.$$

Thus, the use of $T_j[u_n-T_k(u)]$ in (2.1) implies that

$$J_{\Omega,n}\leq c_4\left(\int_{\Omega}f_n\ T_j[u_n-T_k(u)]-\int_{\Omega}\{a(x,u_n,\nabla T_k(u))+\Phi_n(u_n)\}\nabla T_j[u_n-T_k(u)]\right)^{\theta}$$

$$+\ c_3\ meas\{x\in\Omega:|u_n(x)-T_k(u(x))|>j\}^{1-\frac{1}{r}}. \quad (2.12)$$

We remark that

$$\lim_{n\to\infty}\int_{\Omega}f_n\ T_j[u_n-T_k(u)]=\int_{\Omega}f\ T_j[u-T_k(u)]\ =\ \omega_2(k);$$

for $n>j+k$ we have

$$\int_{\Omega}\Phi_n(u_n)\ \nabla T_j[u_n-T_k(u)]\ =\int_{\Omega}\Phi_{j+k}(u_n)\ \nabla T_j[u_n-T_k(u)]\to$$

$$\int_{\Omega}\Phi_{j+k}(u)\ \nabla T_j[u-T_k(u)]\ =\ 0$$

(see ([BDGM], pg. 226); and also

$$\lim_{n\to\infty}\int_{\Omega}a(x,u_n,\nabla T_k(u))\nabla T_j[u_n-T_k(u)]=\int_{\Omega}a(x,u,\nabla T_k(u))\nabla T_j[u-T_k(u)]\ =0;$$

$$\limsup_{n\to\infty} meas\{|u_n(x) - T_k(u(x))| > j\}^{1-\frac{1}{r}} \le meas\{|u(x) - T_k(u(x))| \ge j\}^{1-\frac{1}{r}} =$$

$$= \omega_3(k).$$

Thus

$$\lim_{n\to\infty} J_{\Omega,n} \le c_1\omega_2(k)^\theta + c_3\,\omega_3(k).$$

Hence we have proved that

$$\lim_{n\to\infty} [I_{C_k,n} + I_{A_k,n}] \le \omega_1(k) + c_1\omega_2(k)^\theta + c_3\omega_3(k).$$

Therefore

$$\int_\Omega \{[a(x,u_n,\nabla u_n) - a(x,u_n,\nabla u)]\nabla(u_n - u)\}^\theta \ \to\ 0,$$

that is

$$\|\{[a(x,u_n,\nabla u_n) - a(x,u_n,\nabla u)]\nabla(u_n - u)\}^\theta\|_{L^1(\Omega)} \ \to\ 0,$$

which implies (for a suitable subsequence, still denoted by $u_n$)

$$\{[a(x,u_n,\nabla u_n) - a(x,u_n,\nabla u)]\nabla(u_n - u)\}^\theta \ \to\ 0 \quad a.e.,$$

and also (since $\theta$ is positive)

$$\{[a(x,u_n,\nabla u_n) - a(x,u_n,\nabla u)]\nabla(u_n - u)\} \ \to\ 0 \quad a.e..$$

In [LL], it is proved that, under our assumptions on the function $a(x,s,\xi)$, the previous limit implies that

$$\nabla u_n(x) \ \to\ \nabla u(x) \qquad a.e.. \tag{2.13}$$

The *a priori* estimate of Step 1 yields, for any measurable subset $E \subset \Omega$ that,

$$\lim_{meas(E)\to 0} \int_E |\nabla u_n|^s \ = 0 \qquad \forall s < \tilde{p}, \qquad \text{uniformly with respect to n,} \tag{2.14}$$

since

$$\int_E |\nabla u_n|^s \le \left(\int_\Omega |\nabla u_n|^s\right)^{\frac{s}{q}} meas(E)^{1-\frac{s}{q}} \le c_6 meas(E)^{1-\frac{s}{q}}.$$

Thus, using Vitali Theorem, combined with (2.13) and (2.14), we have

$$\nabla u_n \to \nabla u, \qquad \text{strongly in } (L^s(\Omega))^N, \qquad \forall s < \tilde{p}, \tag{2.15}$$

$$T_k(u_n) \to T_k(u) \qquad \text{strongly in } W_0^{1,\sigma}(\Omega)\ \forall k \geq 0\,,\ \forall \sigma < p. \tag{2.16}$$

**Step 4 - Passage to the limit**

*(Molto allegro)*

Combining (2.15) and the assumptions on the function $a(x,s,\xi)$, we deduce that

$$a(x,u_n,\nabla u_n) \;\to\; a(x,u,\nabla u) \qquad \text{in} \quad L^\rho(\Omega) \quad \forall \rho < \frac{N}{N-1}.$$

Moreover the convergence (2.13), (2.15), and the Fatou Lemma imply that

$$\liminf_{n\to\infty} \int_\Omega a(x,u_n,\nabla u_n)\nabla T_k[u_n-\varphi]$$

$$\geq \liminf_{n\to\infty} \int_\Omega [a(x,u_n,\nabla u_n) - a(x,u_n,\nabla\varphi)]\nabla T_k[u_n-\varphi]$$

$$+ \lim \int_\Omega a(x, T_{k+\|\varphi\|_{L^\infty(\Omega)}}(u_n), \nabla\varphi)\nabla T_k[u_n-\varphi]$$

$$\geq \int_\Omega [a(x,u,\nabla u) - a(x,u,\nabla\varphi)]\nabla T_k[u-\varphi] \;+\; \int_\Omega a(x,u,\nabla\varphi)\nabla T_k[u-\varphi]$$

$$= \int_\Omega a(x,u,\nabla u)\nabla T_k[u-\varphi]$$

Furthermore, for $n \geq k + \|\varphi\|_{L^\infty(\Omega)}$,

$$\int_\Omega \Phi_n(u_n)\nabla T_k[u_n-\varphi] \;=\; \int_\Omega \Phi_n(T_{k+\|\varphi\|_{L^\infty}}(u_n))\nabla T_k[u_n-\varphi]$$

$$= \int_\Omega \Phi(T_{k+\|\varphi\|_{L^\infty}}(u_n))\nabla T_k[u_n-\varphi] \;\to\; \int_\Omega \Phi(u)\nabla T_k[u-\varphi],$$

and

$$\lim_{n\to\infty} \int_\Omega f_n T_k[u_n-\varphi] \;=\; \int_\Omega f T_k[u-\varphi]$$

Then using $T_k[u_n-\varphi]$ as test function in (2.1) and we passing to the limit thanks to the previous results, we prove the existence of a solution $u$ of the nonlinear Dirichlet problem (2.3).

**Remark 2.3.** In order to drop the restriction $p \leq 2 - \frac{1}{N}$ we refer to [BBGV].

**Remark 2.4.** Observe that, in the proof of the previous theorem, we used only the weak $L^1$-convergence of the sequence $f_n$

**Remark 2.5.** The result of the previous theorem still holds if the right hand side is $f - div(F)$, with $F \in (L^{p'}(\Omega))^N$ (see also [BGO]).

**Remark 2.6.** In the previous Theorem 2.1, an important role is played by the truncation operator $T_j$, ($j \in \mathbf{N}$). The method can be adapted easily to the case of right hand side measure taking $j$ arbitrarily <u>small</u>, in order to prove the following theorem.

**Theorem 2.7.** *Let $p$ be such that $2 - \frac{1}{N} < p < N$. Let $A$ be as before, $\Phi = 0$ and $\mu$ be a bounded measure on $\Omega$. Let $\{f_n\}$ be any sequence of smooth function that converges to $\mu$ in the weak sense of the measures and such that $\|f_n\|_{L^1(\Omega)} \leq \|\mu\|_{\mathcal{M}(\Omega)}$, and let $u_n$ be a solution of the boundary value problem*

$$\begin{cases} A(u_n) = f_n & \text{in } \Omega, \\ u_n = 0 & \text{on } \partial\Omega. \end{cases} \tag{2.17}$$

*Then for every $q < \tilde{p} = \frac{N(p-1)}{N-1}$ there exists a positive constant $c_q$ (independent of $n$) such that*

$$\|u_n\|_{W_0^{1,q}(\Omega)} \leq c_q \qquad \forall n \in \mathbf{N}. \tag{2.18}$$

*Moreover, $u_n$ converges in $W_0^{1,q}(\Omega)$ to a function $u$ that is a weak solution of the Dirichlet problem*

$$u \in W_0^{1,q}(\Omega) \ \forall q < \tilde{p} : A(u) = \ \mu . \tag{2.19}$$

## 3. QUALITATIVE PROPERTIES OF THE ENTROPY SOLUTIONS

In this section we study positivity and stability of the entropy solutions depending on positivity and weak-$L^1$ perturbation of the right hand side. For continuous dependence of entropy solutions with respect to perturbation of the differential operator with respect to the $G$-convergence see [B2].

**Theorem 3.1.** *Let $p$ be such that $2 - \frac{1}{N} < p < N$. Let $A$ be as before, $\{f_n\}$ be any sequence of $L^1(\Omega)$ functions that converges to $f$ weakly in $L^1(\Omega)$ and let $u_n$ be, for every $n$ in $\mathbf{N}$, an entropy solution of the boundary value problem*

$$\begin{cases} A(u_n) = f_n - div\, \Phi(u_n) & \text{in } \Omega, \\ u_n = 0 & \text{on } \partial\Omega. \end{cases} \tag{3.1}$$

*Then for every $q < \tilde{p} = \frac{N(p-1)}{N-1}$ there exists a positive constant $c_q$ (independent of $n$) such that*

$$\|u_n\|_{W_0^{1,q}(\Omega)} \leq c_q \qquad \forall n \in \mathbf{N}. \tag{3.2}$$

*Thus, for some $u \in W_0^{1,q}(\Omega)$, $u_n$ converges, up to some subsequence, to $u$ in $W_0^{1,q}(\Omega)$. Furthermore, $u$ is an entropy solution of the Dirichlet problem*

$$u \in W_0^{1,q}(\Omega)\ \forall q < \tilde{p} : A(u) = f - div\, \Phi(u). \tag{3.3}$$

**Proof.** The sequence $u_n$ is bounded in $W_0^{1,q}(\Omega)$ for any $q < \tilde{p}$. Indeed, taking $k = 1$ and $\varphi = T_h(u_n)$ in the definition of entropy solution, we have

$$\alpha \int_{B_h} |\nabla u_n|^p \leq \|f_n\|_{L^1(\Omega)}, \tag{3.4}$$

where

$$B_h = \{x \in \Omega : h \leq |u_n(x)| < h + 1\}. \tag{3.5}$$

Since (3.4) is exactly the inequality (2.5), we can follow the proof of Step 2 of Theorem 2.1 in order to obtain an *a priori* estimate of the sequence $u_n$ in the Sobolev space $W_0^{1,q}(\Omega), q < \tilde{p}$. Hence, there exists a function $u \in W_0^{1,q}(\Omega)$ (for every $q$ such that $1 \leq q < \tilde{p}$) and a subsequence, still denoted by $u_n$, such that

$$u_n \to u \qquad \text{weakly in } W_0^{1,q}(\Omega),\ \forall q < \tilde{p}. \tag{3.6}$$

Moreover, if we take $\varphi = 0$ in the definition of entropy solution, we prove that the sequence $T_k(u_n)$ is bounded in the Sobolev space $W_0^{1,p}(\Omega)$ and so

$$T_k(u_n) \to T_k(u) \qquad \text{weakly in } W_0^{1,p}(\Omega),\ \forall k \geq 0. \tag{3.7}$$

Now the proof follows exactly Step 3 and 4 of Theorem 2.1. Remark only that now the functions $u_n$ do not belong to $W_0^{1,p}(\Omega)$, but only to $W_0^{1,q}(\Omega)$. Indeed, the proof is based on the study of

$$\int_\Omega \{[a(x,u_n,\nabla u_n) - a(x,u_n,\nabla u)]\nabla(u_n - u)\}^\theta = I_{\Omega,n}$$

which is still well defined, and a possibility of a passage to the limit in

$$\int_\Omega f_n T_k[u_n - \varphi]$$

which is possible by Egorov Theorem.

**Theorem 3.2.** *Let $p$ be such that $2 - \frac{1}{N} < p < N$. Let $A$ be as before, let $f \in L^1(\Omega)$, $f \geq 0$ and let $u$ be an entropy solution of the boundary value problem*

$$\begin{cases} A(u) = f - div\, \Phi(u) & in\ \Omega, \\ u = 0 & on\ \partial\Omega. \end{cases}$$

*Then $u \geq 0$.*

**Proof.** Use $\varphi = T_h(u^+)$ in the definition of entropy solution and remark that:

$$\int_\Omega f T_k[u - T_h(u^+)] \leq \int_{\{u \geq h\}} f T_k[u - T_h(u)],$$

since $f$ is positive and

$$\int_{\{u \geq h\}} f T_k[u - T_h(u)] \to 0 \quad (h \to \infty);$$

$$\int_{\{x:u \geq h\}} |\nabla T_k(u)|^p \to 0 \quad (h \to \infty),$$

since $|\nabla T_k[u]|^p \in L^1(\Omega)$;

$$\int_\Omega \Phi(u) \nabla T_k[u - T_h(u^+)] = 0.$$

Then

$$\int_{\{u<0\}} |\nabla u|^p = 0.$$

## 4. REGULARITY OF SOLUTIONS

In this section we present some summabitity results of the entropy solutions (of the boundary value problem (2.3)) and their gradients.

Taking $k = 1$ and $\varphi = T_h(u)$ in the definition of entropy solution, we have the inequality

$$\alpha \int_{B_h} |\nabla u|^p \leq \int_{\{|u(x)|>h\}} |f|. \tag{4.1}$$

Then, following the proofs of [BG2], inequality (4.1) can be used to prove the regularity theorems below.

**Theorem 4.1.** *If*

$$\int_\Omega |f| \log(1 + |f|) < \infty,$$

*then any entropy solution of (2.3) belongs to* $W_0^{1,\frac{(p-1)N}{N-1}}(\Omega)$.

**Theorem 4.2.** *If* $f \in L^m(\Omega), 1 < m \leq (p^*)'$, *then any entropy solution of (2.3) belongs to* $W_0^{1,(p-1)m^*}(\Omega)$.

Remark that, if $m = (p^*)'$, then $(p-1)m^* = p$. Thus, if $m \geq (p^*)'$, entropy solutions have finite energy.

Furthermore, following the proofs of [BG], we can prove the regularity theorem below.

**Theorem 4.3.** *If* $f \in L^m(\Omega), m \geq (p^*)'$, *then any entropy solution of (2.3) is a (standard) weak solution which belongs to* $W_0^{1,p}(\Omega)$. *Moreover we can state its summability. If* $f \in L^m(\Omega), (p^*)' < m < \frac{N}{p}$, *then* $u \in L^s(\Omega)$, *with* $s = \frac{(pm)^*}{p'}$. *If* $f \in L^m(\Omega), m > \frac{N}{p}$, *then* $u$ *is bounded.*

**Acknowledgement** I wish to thank Andrea Dall'Aglio and Luigi Orsina for scientific help. This work (partially supported by MURST [40%, 60%] research funds) contains the unpublished part of my lecture at the Conference, where I also spoke about the results of [BG1], [BG2], [BG3], [BGV], [BBGV], [BDGM], [Da], [Dv], [O].

## REFERENCES


[BBGV] Ph. Benilan, L. Boccardo, T. Gallouët, R. Gariepy. M. Pierre, J.L. Vazquez, *An $L^1$ theory of existence and uniqueness of solutions for nonlinear elliptic equations*, Ann. Scuola Norm. Sup. Pisa, **22** n. 2 (1995), 240–273.

[B 1] L. Boccardo, *The role of truncates in nonlinear Dirichlet problems in $L^1$*, Proceedings of the Fes Conference (may 1994), Longman, to appear.

[B 2] L. Boccardo, *Homogenization and continuous dependence for Dirichlet problems in $L^1$*, Proceedings of the Conference at Scuola Normale Superiore Pisa (december 1994), to appear.

[BD] L. Boccardo, B. Dacorogna, *Monotonicity of certain differential operators in divergence form*, Manuscripta Math., **64** (1989), 253–260.

[BDGM] L. Boccardo, J. Diaz, D.Giachetti, *Existence and regularity of renormalized solutions for some elliptic problems involving derivatives of nonlinear terms*, J. Diff. Eq., **106** (1993), 215–237.

[BG 1] L. Boccardo, T. Gallouët, *Nonlinear elliptic and parabolic equations involving measure data*, J. Funct. Anal., **87** (1989), 149–169.

[BG 2] L. Boccardo, T. Gallouët, *Nonlinear elliptic equations with right hand side measures*, Comm. Partial Differential Equations, **17** (1992) 641–655.

[BG 3] L. Boccardo, T. Gallouët, *Strongly nonlinear elliptic equations having natural growth terms and $L^1$ data*, Nonlinear Anal., **19** (1992), 573–579.

[BG 4] L. Boccardo, T. Gallouët, *Summability of the solutions of nonlinear elliptic equations with right hand side measures*, Convex Anal., (1996) to appear.

[BG 5] L. Boccardo, T. Gallouet, *Problèmes unilatéraux in $L^1$*, C. R. Acad. Sci. Paris, **311** (1990), pp. 617–619.

[BGM] L. Boccardo, T. Gallouët, F. Murat, *A unified presentation of two existence results for problems with natural growth*, in *Progress in PDE, the Metz surveys 2*, M. Chipot editor, Research Notes in Mathematics, Longman, **296** (1993), 127–137.

[BGO] L. Boccardo, T. Gallouët, L. Orsina, *Existence and uniqueness of entropy solutions for nonlinear elliptic equations with right hand side measures*, Annales IHP, Analyse non Lineaire, to appear.

[BGV] L. Boccardo, T. Gallouët, J.L. Vazquez, *Nonlinear elliptic equations in $\mathbf{R}^N$ without growth restrictions on the data*, J. Differential Equations, **105** (1993), 334–363.

[BGi] L. Boccardo, D. Giachetti, *Existence results via regularity for some nonlinear elliptic problem*, Comm. P.D.E. **14** (1989), 663–680.

[BM] L. Boccardo, F. Murat, *Almost everywhere convergence of the gradients of solutions to elliptic and parabolic equations*, Nonlinear Anal. TMA **19** (1992), 581–597.

[BS] H. Brezis, W. Strauss, *Semilinear elliptic equations in $L^1$*, J. Math. Soc. Japan **25** (1973), 565–590.

[Da] A. Dall'Aglio, *Approximated solutions of equations with $L^1$ data. Application to the $H$-convergence of quasi-linear parabolic equations*, Ann. Mat. Pura Appl., to appear.

[Dv] T. Del Vecchio, *Nonlinear elliptic equations with measure data*, Potential Analysis, **4**(1995), 185–203.

[LL] J. Leray, J.L. Lions, *Quelques résultats de Višik sur les problèmes elliptiques semi-linéaires par les méthodes de Minty et Browder*, Bull. Soc. Math. France, **93** (1965) 97–107.

[LM] P.L. Lions, F. Murat, *Sur les solutions renormalisées d'équations elliptiques non linéaires*, to appear.

[M] F. Murat, *Soluciones renormalizadas de EPD no lineales*, preprint Université Paris VI (1993).

[O] L. Orsina, *Solvability of linear and semilinear eigenvalue problems with $L^1$ data*, Rend. Sem. Mat. Univ. Padova, **90** (1993), 207–238.

[P] A. Prignet, *Remarks on existence and uniqueness of solutions of elliptic problems with right hand side measures*, Rend Mat., **15** (1995), 321–337.

[Se] J. Serrin, *Pathological solutions of elliptic differential equations*, Ann. Scuola Norm. Sup. Pisa, **18** (1964), 385–387.

[St] G. Stampacchia, *Équations elliptiques du second ordre à coefficients discontinus* Les Presses de l'Université de Montréal, Montréal (1966).

Lucio Boccardo

Dipartimento di Matematica, Università di Roma I

Piazza A. Moro 2, 00185, Roma, Italia

tel: (39.6)49913202; e-mail: boccardo@gpxrme.sci.uniroma1.it

G. BUTTAZZO

# A general theory of relaxed controls and application to shape optimization problems

## 1. Introduction

A shape optimization problem is a problem of the form

$$\min\{J(A) \ : \ A \in \mathcal{A}\} \tag{1.1}$$

where $\mathcal{A}$ is the class of all *admissible domains*, and $J$ is the *cost function* which has to be minimized. In a large number of situations the cost $J$ depends on the domain variable $A$ through the solution $u_A$ of a partial differential equation in $A$ which can be considered as the *state equation* of an optimal control problem. More precisely, if the control variables are the domains of $\mathcal{A}$, and the state equation $E$ associates to every $A \in \mathcal{A}$ the solution $u_A$ of a PDE in $A$ of the form $u = E(A)$, the optimal control problem can be written as

$$\min\{J(A, u_A) \ : \ A \in \mathcal{A}\} \tag{1.2}$$

or equivalently, if $U$ denotes the space of states, as

$$\min\{J(A, u) \ : \ A \in \mathcal{A},\ u \in U,\ u = E(A)\}. \tag{1.3}$$

Note that here the cost function $J$ may depend both on $A$ and $u_A$.

Problems of this form are very common in a lot of applications to structural engineering, where $A$ can be the design of a mechanical structure, $u$ its displacement, and the cost $J$ may take into account the weight of the structure and some desired performances given in terms of the displacement $u$.

Here we simply consider the case when the state equation is an elliptic equation with Dirichlet boundary conditions, and we take $-\Delta$ (on a given subset $\Omega$) as a model operator, referring to Buttazzo and Dal Maso [7], [8] for further generalizations. As for the cost functional, we consider the case of an integral either of the form

$$J(u) = \int_\Omega j(x,u)\,dx \tag{1.4}$$

in which there is no explicit dependence on the control variable $A$, or

$$J(A,u) = \int_A j(x,u)\,dx. \tag{1.5}$$

The class $\mathcal{A}$ of all admissible domains is taken as the class of all subdomains of $\Omega$. Other kinds of shape optimization problems with different boundary conditions on the free interface have been widely considered in the literature; we refer for instance to the papers by Kohn and Strang [17], and by Murat and Tartar [19].

When one considers the existence problems for (1.1) the main difficulty consists in the fact that the class $\mathcal{A}$ does not have a standard topology, suitable to apply the direct method of the calculus of variations; in fact, in general, we do not have to expect problem (1.1) has a solution. The same situation may occur in other kinds of optimal control problems where the set of admissible controls do not have a topology at all or has a topology which is not suitable for the problem we are dealing with. For this reason, in Belloni, Buttazzo and Freddi [1] we developed a general framework for relaxed optimal control problems: we define a class $\hat{\mathcal{A}}$ of *relaxed controls*, and give an existence result for minimizers in $\hat{\mathcal{A}}$.

Here we recall the definition of relaxed controls, together with their main properties in an abstract setting; then we show how the shape optimization problem introduced above can be described in terms of relaxation. Finally, we recall a result by Buttazzo and Dal Maso [8] in which, under particular assumptions on the cost $J$, it is possible to prove that problem (1.1) admits a solution in the class $\mathcal{A}$ of original domains, and we give some related examples.

## 2. Relaxed Controls

In this section we define in a very general setting the class of relaxed controls. Consider an abstract optimal control problem where

$U$ is the space of states, we assume it is a topological space;

$\mathcal{A}$ is the set of controls, no topology on $\mathcal{A}$ is a priori considered;

$J$ is the cost function defined on $\mathcal{A} \times U$;

$G$ is a function defined on $\mathcal{A} \times U$ which provides the state equation as a minimization problem of the form

$$u \in \operatorname{argmin} G(A, \cdot);$$

we call $G$ the *state functional.*

Therefore, the optimal control problem is written as

$$(P) \qquad \min \{ J(A, u) \ : \ A \in \mathcal{A}, \ u \in U, \ u \in \operatorname{argmin} G(A, \cdot) \}.$$

We make the following assumptions:

(2.1) the space $U$ of states is a separable metric space;

(2.2) for every $A \in \mathcal{A}$ the function $G(A, \cdot)$ is proper and lower semicontinuous on $U$;

(2.3) there exists a lower semicontinuous function $\psi : U \to \overline{\mathbf{R}}$ which is coercive, in the sense that for every $t \in \mathbf{R}$ the sublevel set $\{\psi \leq t\}$ is compact in $U$, such that

$$G(A, u) \geq \psi(u) \qquad \forall (A, u) \in \mathcal{A} \times U.$$

It is well known (see for instance Dal Maso [12], Chapter 10) that under the assumptions above the class $S_\psi(U)$ of all lower semicontinuous functions from $U$ into $\overline{\mathbf{R}}$ minorized by $\psi$, endowed by the $\Gamma$-convergence becomes a compact metric space with a distance $\delta_\Gamma$. Therefore, if we assume

(2.4) the mapping $\mathcal{G} : \mathcal{A} \to S_\psi(U)$ given by $\mathcal{G}(A) = G(A, \cdot)$ is one-to-one

and we set

$$d(A, B) = \delta_\Gamma(\mathcal{G}(A), \mathcal{G}(B)),$$

we get that $d$ is a distance on $\mathcal{A}$ and $\mathcal{G}$ is an isometry between $(\mathcal{A}, d)$ and $(S_\psi(U), \delta_\Gamma)$, so that

$$A_h \to A \text{ in } \mathcal{A} \quad \Longleftrightarrow \quad \Gamma \lim_{h\to+\infty} G(A_h, \cdot) = G(A, \cdot).$$

Since the $\Gamma$-convergence implies the convergence of minima and of minimum points (see for instance Dal Maso [12]), we have topologized $\mathcal{A}$ in the most natural way with respect to the minimum problem $(P)$. We are now in a position to introduce the relaxed controls and to define the relaxed problem $(\overline{P})$.

Let $(\hat{\mathcal{A}}, \hat{d})$ be the completion of the metric space $(\mathcal{A}, d)$ and let $\hat{\mathcal{G}} : \hat{\mathcal{A}} \to \overline{\mathbf{R}}$ be the unique isometry which extends $\mathcal{G}$; more precisely we set

$$\hat{\mathcal{G}}(\hat{A}) = \Gamma \lim_{h\to+\infty} \mathcal{G}(A_h)$$

where $(A_h)$ is any sequence converging to $\hat{A}$ in $(\hat{\mathcal{A}}, \hat{d})$. In other words, if we define $\hat{G} : \hat{\mathcal{A}} \times U \to \overline{\mathbf{R}}$ by $\hat{G}(\hat{A}, \cdot) = \hat{\mathcal{G}}(\hat{A})$, we have

$$\hat{A}_h \to \hat{A} \text{ in } (\hat{\mathcal{A}}, \hat{d}) \quad \Longleftrightarrow \quad \Gamma \lim_{h\to+\infty} \hat{G}(\hat{A}_h, \cdot) = \hat{G}(\hat{A}, \cdot).$$

By the compactness of the $\Gamma$-convergence on $S_\psi(U)$ we get that $(\hat{\mathcal{A}}, \hat{d})$ is a compact metric space. The elements of $\hat{\mathcal{A}}$ are called *relaxed controls* and $\hat{G}$ is the *relaxed state functional.* In order to identify the relaxed control problem $(\overline{P})$ assume

(2.5) for every $\hat{A} \in \hat{\mathcal{A}}$ the minimum point of $\hat{G}(\hat{A}, \cdot)$ on $U$ is unique;

(2.6) there exists a function $\Phi : \mathcal{A} \to \mathbf{R}$ bounded on $d$-bounded sets, and a function $\omega : U \times U \to \mathbf{R}$ with

$$\lim_{v\to u} \omega(u, v) = 0 \qquad \forall u \in U,$$

such that

$$|J(A, u) - J(A, v)| \le \Phi(A)\omega(u, v) \qquad \forall A \in \mathcal{A},\ \forall u, v \in U.$$

**Theorem 2.1.** (see Belloni, Buttazzo, Freddi [1]) *Under assumptions (2.1)–(2.6) the relaxed problem associated to (P) is given by*

$$(\overline{P}) \qquad \min\left\{\hat{J}(\hat{A}, u) \ :\ \hat{A} \in \hat{\mathcal{A}},\ u \in U,\ u \in \operatorname{argmin} \hat{G}(\hat{A}, \cdot)\right\}$$

*where $\hat{J} : \hat{\mathcal{A}} \times U \to \overline{\mathbf{R}}$ is the relaxed cost functional*

$$\hat{J}(\hat{A}, u) = \inf\left\{\liminf_{h\to+\infty} J(A_h, u) \ : \ A_h \to \hat{A} \text{ in } \hat{\mathcal{A}}\right\}.$$

*Problem $(\overline{P})$ admits always a solution, the minimizing sequences of $(P)$ converge (up to subsequences) to minimum points of $(\overline{P})$, and*

$$\inf P = \min \overline{P}.$$

**Remark 2.2.** When the cost functional $J$ does not depend on the control variable $A$, we do not need to perform any relaxation in the cost; indeed problem $(\overline{P})$ is simply

$$(\overline{P}) \qquad \min\left\{J(u) \ : \ \hat{A} \in \hat{\mathcal{A}}, \ u \in U, \ u \in \operatorname{argmin} \hat{G}(\hat{A}, \cdot)\right\}.$$

## 3. Relaxed Shape Optimization Problems

In this section we consider a class of shape optimization problems and we identify their relaxed formulation. Consider a bounded open subset $\Omega$ of $\mathbf{R}^n$ $(n \geq 2)$ with a Lipschitz boundary, and let $f \in H^{-1}(\Omega)$ be given. With the notation of Section 2 we have

$U$ is the space $H_0^1(\Omega)$ endowed with the strong $L^2(\Omega)$ topology;

$\mathcal{A}$ is the class of all open subsets $A$ of $\Omega$;

$G$ is the state functional which provides the elliptic equation

$$\begin{cases} -\Delta u = f & \text{in } A \\ u = 0 & \text{on } \Omega \setminus A, \end{cases}$$

that is

$$G(A, u) = \begin{cases} \int_\Omega |Du|^2\,dx - 2\langle f, u\rangle & \text{if } u \in H_0^1(A) \\ +\infty & \text{otherwise;} \end{cases}$$

$J$ is an integral cost functional of the form (1.4) or (1.5), where $j(x,s)$ is a Carathéodory integrand satisfying

$$|j(x,s)| \le a(x) + b|s|^2$$

for suitable $a \in L^1(\Omega)$ and $b \ge 0$.

The characterization of the space $\hat{\mathcal{A}}$ of relaxed controls has been obtained by Dal Maso and Mosco in [14] where it is proved that $\hat{\mathcal{A}}$ coincides with the class $\mathcal{M}_0$ of all Borel measures $\mu : \mathcal{B}(\Omega) \to [0, +\infty]$ which vanish on all sets of capacity zero. The relaxed cost functional $\hat{G}$ is then given by

$$\hat{G}(\mu, u) = \int_\Omega |Du|^2\,dx - 2\langle f, u\rangle + \int_\Omega u^2\,d\mu \tag{3.1}$$

so that the relaxed state equation $u \in \operatorname{argmin} \hat{G}(\mu, \cdot)$ can be written as

$$\begin{cases} -\Delta u + \mu u = f & \text{in } \Omega \\ u \in H_0^1(\Omega) \cap L_\mu^2(\Omega). \end{cases} \tag{4.2}$$

Note that, being $u \in H_0^1(\Omega)$, the value $u(x)$ is defined almost everywhere in capacity, so that the integral $\int_\Omega u^2\,d\mu$ is well defined. Moreover, when

$$\mu(B) = \infty_{\Omega\setminus A}(B) = \begin{cases} 0 & \text{if } \operatorname{cap}(A \setminus B) = 0 \\ +\infty & \text{otherwise,} \end{cases}$$

it is easy to see that $\hat{G}(\mu, \cdot) = G(A, \cdot)$ and equation (4.2) reduces to the original equation

$$\begin{cases} -\Delta u = f & \text{in } A \\ u = 0 & \text{on } \Omega \setminus A. \end{cases}$$

It remains to characterize the relaxed cost functional $\hat{J}$. If $J$ is of the form (1.4) without any explicit dependance on $A$, then $\hat{J} = J$ and so the relaxed shape optimization problem reads

$$(\overline{P}) \qquad \min\left\{\int_\Omega j(x,u)\,dx \;:\; \mu \in \mathcal{M}_0,\ u \in H_0^1(\omega),\ -\Delta u + \mu u = f\right\}.$$

When $J$ is of the form (1.5) Buttazzo, Dal Maso, Garroni and Malusa [9] obtained the following expression for $\hat{J}$:

$$\hat{J}(\mu, u) = \int_{A_\mu} j(x,u)\,dx - \int_{\Omega\setminus A_\mu} j^-(x,0)\,dx$$

being, roughly speaking, $A_\mu$ the set where $\mu$ "is finite". More precisely, if for every $\mu \in \mathcal{M}_0$ we denote by $w_\mu$ the solution of

$$\begin{cases} -\Delta w + \mu w = 1 \\ w \in H_0^1(\Omega) \cap L_\mu^2(\Omega) \end{cases}$$

then $A_\mu$ coincides with the set $\{w_\mu > 0\}$.

In a similar way (see [9]) we can treat cost functionals of the form

$$J(A,u) = \int_A j(x,u)\,d\lambda \tag{3.3}$$

where $\lambda$ is a given finite measure of $\mathcal{M}_0$ and $j(x,s)$ is an integrand such that

(i) for every $u \in H_0^1(\Omega)$ the function $j\big(x,u(x)\big)$ belongs to $L_\lambda^1(\Omega)$;

(ii) the map $u \mapsto \int_\Omega j(x,u)\,d\lambda$ is sequentially weakly continuous in $H_0^1(\Omega)$.

In this case we get

$$\hat{J}(\mu,u) = \int_{A_\mu} j(x,u)\,d\lambda - \int_{\Omega\setminus A_\mu} j^-(x,0)\,d\lambda.$$

The case when a constraint of the form $|A| \in T$ is present in the original shape optimization problem $(P)$, is slightly more complicated; here $T$ is any subset of the interval $[0,|\Omega|]$ and $|\cdot|$ denotes the Lebesgue measure (or the measure $\lambda$ if $J$ is given by (3.3)). Then we obtain (see [9]) for $\hat{J}$ the expression

$$\hat{J}(\mu,u) = \begin{cases} \int_\Omega j(x,u)\,dx & \text{if } |A_\mu| \le \sup T \\ +\infty & \text{otherwise} \end{cases}$$

if $J$ is of the form (1.4), and

$$\hat{J}(\mu,u) = \int_{A_\mu} j(x,u)\,dx + \inf\left\{\int_{B\setminus A_\mu} j(x,0)\,dx \ :\ A_\mu \subset B,\ |B| \in T\right\}$$

if $J$ is of the form (1.5).

## 4. Existence of Nonrelaxed Solutions

In this section we consider particular assumptions on the cost $J$ yielding to the existence of an optimal domain for the problem

$$\min\{J(A) \ : \ A \in \mathcal{A}\}. \tag{4.1}$$

An easy situation is when the admissible domains of the class $\mathcal{A}$ have a Lipschitz boundary whose Lipschitz constant is a priori bounded from above; in this case, by using the direct method of the calculus of variations, it is possible to prove the existence of a solution of problem (4.1) under very mild assumptions on the cost functional $J$ (see for instance Pironneau [20], Chapter 3, Theorem 2). However, constraints of this kind are sometimes unnatural, whereas volume constraints like $|A| = c$ seem more appropriate.

Given a bounded open subset $\Omega$ of $\mathbf{R}^n$, let us take the class $\mathcal{A}$ of admissible domains for (4.1) as the class of all *quasi-open* subsets $A$ of $\Omega$, where quasi-open means that for every $\varepsilon > 0$ there exists an open set $A_\varepsilon$ such that $\text{cap}(A \Delta A_\varepsilon) < \varepsilon$.

For every quasi-open subset $A$ of $\Omega$ the space $H_0^1(A)$ is defined as the class of all functions $u \in H_0^1(\Omega)$ such that $u = 0$ cap-a.e. on $\Omega \setminus A$. In this way $H_0^1(A)$ is a closed subspace of $H_0^1(\Omega)$, from which it inherits the Hilbert space structure.

The class $\mathcal{A}$ can be endowed with the distance $d$ described in the previous sections; in other words we have $A_h \to A$ if and only if (see Dal Maso [12], Theorem 13.12) for every $f \in H^{-1}(\Omega)$ the solutions $u_h$ of

$$\begin{cases} -\Delta u = f & \text{in } A_h \\ u \in H_0^1(A) \end{cases} \tag{4.2}$$

converge strongly in $H_0^1(\Omega)$ to the solution $u$ of

$$\begin{cases} -\Delta u = f & \text{in } A \\ u \in H_0^1(A). \end{cases} \tag{4.3}$$

Equivalently (see Sverak [21], Corollary 3.3) we have $A_h \to A$ in $\mathcal{A}$ if and only if the solutions $u_h$ of (4.2) converge strongly in $H_0^1(\Omega)$ to the solution $u$ of (4.3) only for $f \equiv 1$.

The existence result proved in Buttazzo and Dal Maso [8] is the following.

**Theorem 4.1.** *Let $J : \mathcal{A} \to \overline{\mathbf{R}}$ be a mapping satisfying the following assumptions:*

*(i) $J$ is lower semicontinuous with respect to the convergence in $\mathcal{A}$ introduced above;*

*(ii) $J$ is decreasing with respect to the set inclusion, that is*

$$A \subset B \quad \Rightarrow \quad J(A) \geq J(B).$$

*Then for every $c \in [0, |\Omega|]$ the problem*

$$\min \{ J(A) \ : \ A \in \mathcal{A}, \ |A| = c \} \tag{4.4}$$

*admits at least a solution.*

**Remark 4.2.** Even if assumption (i) in Theorem 4.1 relies on the topology of $\mathcal{A}$ and so it is quite involved, it is satisfied in a large number of situations; on the contrary, assumption (ii) is much stronger, and may fail even in very simple cases as

$$J(A) = \int_\Omega |u_A - \phi(x)|^2 \, dx \tag{4.5}$$

where $u_A$ is the solution of (4.3) with a given $f$. On the other hand, it is possible to see that shape optimization problems with $J$ of the form (4.5) do not have in general domain solutions but only relaxed solutions (see for instance Buttazzo [4], Chipot and Dal Maso [10]).

We present here some relevant examples in which Theorem 4.1 applies.

**Example 4.3.** For every $A \in \mathcal{A}$ let

$$\Lambda(A) = (\lambda_1(A), \lambda_2(A), \ldots)$$

be the sequence of all eigenvalues of $-\Delta$ on $H_0^1(A)$, arranged in a nondecreasing way and repeated according to their multiplicity. We adopt the convenction that $\lambda_k(A) = +\infty$ for every $k \in \mathbf{N}$ if $\mathrm{cap}(A) = 0$. It is well known that for every $k \in \mathbf{N}$ the mapping $\lambda_k : \mathcal{A} \to \overline{\mathbf{R}}$ is continuous for the topology of $\mathcal{A}$ (see for instance Dunford and Schwartz [15], Lemma XI.9.5) and decreasing with respect to the set inclusion (see for instance Courant and Hilbert [11], Chapter VI, Theorem 3). Therefore, by Theorem 4.1, for every mapping $\Phi : \overline{\mathbf{R}}^{\mathbf{N}} \to \overline{\mathbf{R}}$ lower semicontinuous and increasing in the sense that

$$\lambda_k^h \to \lambda_k \quad \forall k \in \mathbf{N} \quad \Rightarrow \quad \Phi(\Lambda) \le \liminf_{h\to+\infty} \Phi(\Lambda^h)$$
$$\lambda_k^1 \le \lambda_k^2 \quad \forall k \in \mathbf{N} \quad \Rightarrow \quad \Phi(\Lambda^1) \le \Phi(\Lambda^2),$$

the minimum problem

$$\min\left\{\Phi\big(\Lambda(A)\big) \;:\; A \in \mathcal{A},\ |A| = c\right\}$$

admits at least a solution. For instance, problems

$$\min\left\{\lambda_1(A) + \lambda_2(A) \;:\; A \in \mathcal{A},\ |A| = c\right\}$$
$$\min\left\{\lambda_1^2(A) + \lambda_2^2(A) \;:\; A \in \mathcal{A},\ |A| = c\right\}$$

both admit a solution.

**Example 4.4.** Let $f \in H^{-1}(\Omega)$ be with $f \ge 0$, and let $g : \Omega \times \mathbf{R} \to \overline{\mathbf{R}}$ be a Borel function such that $g(x, \cdot)$ is lower semicontinuous and decreasing on $\mathbf{R}$ for a.e. $x \in \Omega$, and for suitable $a \in L^1(\Omega)$ and $b \in \mathbf{R}$

$$g(x, s) \ge -a(x) - b|s|^2.$$

If for every $A \in \mathcal{A}$ we denote by $u_A$ the solution of (4.3), and set

$$J(A) = \int_\Omega g\big(x, u_A(x)\big)\, dx,$$

then $J$ turns out to be lower semicontinuous with respect to the topology of $\mathcal{A}$ and, by the maximum principle and the monotonicity of $g(x, \cdot)$, decreasing with respect to the set inclusion. Therefore, by Theorem 4.1 the minimum problem

$$\min\left\{\int_\Omega g(x, u_A)\, dx \;:\; A \in \mathcal{A},\ |A| = c\right\}$$

admits at least a solution. For instance $g(x, s) = -s$ or $g(x, s) = -s|s|$ satisfy the assumptions above.

**Example 4.5.** Consider for every $A \in \mathcal{A}$

$$J(A) = \text{cap}(\Omega \setminus A).$$

It is possible to prove that $J$ is lower semicontinuous with respect to the topology of $\mathcal{A}$, and moreover $J$ is clearly decreasing with respect to the set inclusion. Thus, by Theorem 4.1, we obtain the existence of a solution for the minimum problem

$$\min \left\{ \text{cap}(\Omega \setminus A) \ : \ A \in \mathcal{A}, \ |A| = c \right\}$$

from which it is not difficult to get a solution for

$$\min \left\{ \text{cap}(E) \ : \ E \subset \Omega, \ |E| = k \right\}.$$

**Acknowledgements.** This work is part of the project "EURHomogenization", contract SC1-CT91-0732 of the program SCIENCE of the Commission of the European Communities.

## References

[1] M. BELLONI, G. BUTTAZZO, L. FREDDI: *Completion by Gamma-convergence for optimal control problems.* Ann. Fac. Sci. Toulouse Math., **2** (1993), 149–162.

[2] G. BUTTAZZO: *Semicontinuity, Relaxation and Integral Representation in the Calculus of Variations.* Pitman Res. Notes Math. Ser. **207**, Longman, Harlow (1989).

[3] G. BUTTAZZO: *Relaxed formulation for a class of shape optimization problems.* Proceedings of "Boundary Control and Boundary Variations", Sophia-Antipolis, October 15–17, 1990, Lecture Notes in Control and Inf. Sci. **178**, Springer-Verlag, Berlin (1992), 50–59.

[4] G. BUTTAZZO: *Existence via relaxation for some domain optimization problems.* Proceedings of "Topology Design of Structures", Sesimbra 20–26 June 1992, NATO ASI Series E: Applied Sciences **227**, Kluwer Academic Publishers, Dordrecht (1993), 337–343.

[5] G. BUTTAZZO: *Relaxed shape optimization problems with Dirichlet boundary conditions.* Preprint Dipartimento di Matematica Università di Pisa, Pisa (1994).

[6] G. BUTTAZZO, G. DAL MASO: *Shape optimization for Dirichlet problems: relaxed solutions and optimality conditions.* Bull. Amer. Math. Soc., **23** (1990), 531–535.

[7] G. BUTTAZZO, G. DAL MASO: *Shape optimization for Dirichlet problems: relaxed formulation and optimality conditions.* Appl. Math. Optim., **23** (1991), 17–49.

[8] G. BUTTAZZO, G. DAL MASO: *An existence result for a class of shape optimization problems.* Arch. Rational Mech. Anal., **122** (1993), 183–195.

[9] G. BUTTAZZO, G. DAL MASO, A. GARRONI, A. MALUSA: *On the relaxed formulation of some shape optimizatio problems.* Preprint SISSA, Trieste (1994).

[10] M. CHIPOT, G. DAL MASO: *Relaxed shape optimization: the case of nonnegative data for the Dirichlet problem.* Adv. Math. Sci. Appl., **1** (1992), 47–81.

[11] R. COURANT, D. HILBERT: *Methods of Mathematical Physics.* Whiley, New York (1953).

[12] G. DAL MASO: *An Introduction to Γ-convergence.* Birkhäuser, Boston (1993).

[13] G. DAL MASO, A. MALUSA: *Approximation of relaxed Dirichlet problems by boundary value problems in perforated domains.* Proc. Roy. Soc. Edinburgh, (to appear).

[14] G. DAL MASO, U. MOSCO: *Wiener's criterion and Γ-convergence.* Appl. Math. Optim., **15** (1987), 15–63.

[15] N. DUNFORD, J.T. SCHWARTZ: *Linear Operators.* Interscience, New York (1957).

[16] S. FINZI VITA: *Numerical shape optimization for relaxed Dirichlet problems.* Math. Mod. Meth. Appl. Sci., **3** (1993), 19–34.

[17] R. V. KOHN, G. STRANG: *Optimal design and relaxation of variational problems, I, II, III.* Comm. Pure Appl. Math., **39** (1986), 113–137, 139–182, 353–377.

[18] F. MURAT, J. SIMON: *Sur le contrôle par un domaine géometrique.* Preprint 76015, Univ. Paris VI, (1976).

[19] F. MURAT, L. TARTAR: *Optimality conditions and homogenization.* Proceedings of "Nonlinear variational problems", Isola d'Elba 1983, Res. Notes in Math. **127**, Pitman, London, (1985), 1–8.

[20] O. PIRONNEAU: *Optimal Shape Design for Elliptic Systems.* Springer-Verlag, Berlin (1984).

[21] V. SVERAK: *On optimal shape design.* Preprint Heriot-Watt University, Edinburgh (1992) and C. R. Acad. Sci. Paris, **I-315** (1992), 545–549.

Giuseppe BUTTAZZO
Dipartimento di Matematica
Università di Pisa
Via Buonarroti, 2
56127 PISA (ITALY)

## A. CANFORA

# On the elliptic equations with non $VMO$ coefficients

### 1. Miranda Theorems

To enter upon the subject it is necessary to recall a classical paper of C. Miranda ([M1], 1963) concerning the elliptic equations with $W^{1,n}$ coefficients.

To simplify, we shall expose Miranda method with regard to the "canonical" equation:

$$(1)\qquad \begin{cases} -\sum_{i}^{1\dots n} A_i D_{x_i}^2 u + \lambda u = f\ , & \text{in } G \equiv ]0, 2\pi[^n \\ \\ f \in L^2(G)\ , \quad u \in W_{\#}^{2,2}(G)\ , \end{cases}$$

where: $C_{\#}^{\infty}(\mathbf{R}^n)$ is the class of $C^{\infty}(\mathbf{R}^n)$ functions $2\pi$-periodic in $x_i$ ($i = 1, \dots, n$), and $W_{\#}^{k,p}(G)$ is the closure of $C_{\#}^{\infty}(\mathbf{R}^n)$ in $W^{k,p}(G)$; so the purpose is to find strong periodic solutions of equation (1).

The hypotheses are:

$$(2)\qquad \lambda \text{ (constant)} > 0\ , \quad A_i \in W_{\#}^{1,n}(G) \cap L^{\infty}(G)\ , \quad A_i \geq 1\ .$$

Posed initially: $-\sum_{i}^{1\dots n} A_i D_{x_i}^2 u + \lambda u \equiv Lu = f$ (with $u \in C_{\#}^{\infty}(\mathbf{R}^n)$), multiplying by $-D_{x_j}^2 u$, integrating on $G$ and summing on $j = 1, \dots, n$, one has:

$$\sum_{i,j}^{1\dots n} \int_G A_i D_{x_i}^2 u D_{x_j}^2 u - \lambda \sum_{j}^{1\dots n} \int_G u D_{x_j}^2 u = - \sum_{j}^{1\dots n} \int_G f D_{x_j}^2 u\ ;$$

then , integrating by parts and taking into account the periodicity of $A_1, \dots, A_n$ and $u$, one obtains:

$$(3)\qquad \begin{aligned} |D^2 u|_{2,G}^2 + \lambda |Du|_{2,G}^2 &\leq \sum_{i,j}^{1\dots n} \int_G A_i (D_{x_i} D_{x_j} u)^2 + \\ +\lambda \sum_{j}^{1\dots n} \int_G (D_{x_j} u)^2 &= - \sum_{j}^{1\dots n} \int_G f D_{x_j}^2 u + \\ + \sum_{\substack{i,j \\ i \neq j}}^{1\dots n} \int_G D_{x_j} A_i D_{x_j} u D_{x_i}^2 u &- \sum_{\substack{i,j \\ i \neq j}}^{1\dots n} \int_G D_{x_i} A_i D_{x_j} u D_{x_i} D_{x_j} u \end{aligned}$$

[where $|D^k u|_{p,G}$ are the usual Nirenberg semi-norms]. Analogously, multiplying $Lu = f$ by $u$ and integrating on $G$, one has:

$$(3') \qquad \lambda \int_G u^2 = \int_G fu + \sum_i^{1\dots n} \int_G u A_i D^2_{x_i} u \, .$$

The principal obstacle is the majorization of the integrals in second and third sum of the last member of (3). Let us see, for example, the integrals of second sum. Being $A_i \in W^{1,n}$, we have:

$$(4) \qquad |\int_G D_{x_j} A_i D_{x_j} u D^2_{x_i} u| \leq |D^2_{x_i} u|_{2,G_\varepsilon} |D_{x_j} u|_{\frac{2n}{n-2},G_\varepsilon} |D_{x_j} A_i|_{n,G_\varepsilon} + \\ + M_\varepsilon |D^2_{x_i} u|_{2,G\setminus G_\varepsilon} |D_{x_j} u|_{2,G\setminus G_\varepsilon}$$

where (De La Vallée Poussin Theorem), fixed $\varepsilon > 0$, we have determined an $M_\varepsilon > 0$ such that:

$$(4') \qquad \begin{cases} \int_{G_\varepsilon} |D_{x_j} A_i|^n < \varepsilon^n \, ; \qquad G_\varepsilon \equiv \{x \in G : |D_{x_j} A_i(x)| > M_\varepsilon\} \\ \\ \forall x \in G\setminus G_\varepsilon \, , \qquad |D_{x_j} A_i(x)| \leq M_\varepsilon \, . \end{cases}$$

From (4) and Sobolev inequalities, we deduce:

$$(5) \qquad |\int_G D_{x_j} A_i D_{x_j} u D^2_{x_i} u| \leq |D^2_{x_i} u|_{2,G} K \|u\|_{W^{2,2}(G)} \cdot \varepsilon + \\ + M_\varepsilon |D^2_{x_i} u|_{2,G} \left\{ \varepsilon' |D^2_{x_j} u|_{2,G} + \frac{K'}{\varepsilon'} |u|_{2,G} \right\} \, .$$

From (3) [and (3′)] and majorizations like (5), if $\varepsilon$ and $\varepsilon'$ are taken sufficiently small (and if $\lambda$ is sufficiently large), one obtains the a priori estimate:

$$(6) \qquad \|u\|_{W^{2,2}(G)} \leq C \|Lu\|_{L^2(G)} \, , \qquad \forall u \in W^{2,2}_{\#}(G) \, .$$

On the basis of (6) one establishes

**Theorem $\mathcal{M}_1$** — *In the hypotheses (2), for all $f \in L^2(G)$, equation (1) has one and only one solution $u \in W^{2,2}_{\#}(G)$.*

Truly Miranda result is more articulated and concerns Dirichlet Problem in $C^2$ domains, in the simple hypothesis: $\lambda \geq 0$, $A_i \in W^{1,n} \cap L^\infty$. But the essential question is the majorization of integrals of type (4); here is the principal effort necessary if one has the objective of reducing (in the hypotheses) the regularity of coefficients. Indeed, in the same Miranda paper [M1] (§ 5) one finds an other remarkable result: an Existence (but not Unicity) Theorem [we refer always to problem (1)] in the hypotheses:

$$(2') \qquad f \in L^\infty(G)\ , \quad A_i \in W^{1,4}_{\#}(G)\ , \quad n > 4\ .$$

If $\{A_{i,m}\}$ is a sequence of functions $\in C^\infty_{\#}(\mathbf{R}^n)$ approximating $A_i$ in $W^{1,4}(G)$ [and analogously: $C^\infty_{\#}(\mathbf{R}^n) \ni f_m \xrightarrow[m]{} f$ in $L^2(G)$, and moreover $\sup_G |f_m| \le \sup_G |f|\ , \ \forall m$], the approximating problem:

$$(1_m) \qquad -\sum_i^{1\ldots n} A_{i,m} D^2_{x_i} u + \lambda u \equiv L_m u = f_m$$

has ($\forall m$) one and only one solution $u_m \in C^\infty_{\#}(\mathbf{R}^n)$.
By the maximum principle one has:

$$(7) \qquad \sup_G |u_m| \le \frac{1}{\lambda} \sup_G |f_m| \le \frac{1}{\lambda} \sup_G |f|\ , \quad \forall m\ .$$

Written (3) for $L_m$ and $u_m$, for the integrals (4) [relative to $L_m$ and $u_m$], by means of (7) and Gagliardo - Nirenberg inequality one obtains the estimate:

$$(4'') \qquad \left| \int_G D_{x_j} A_{i,m} D_{x_j} u_m D^2_{x_i} u_m \right| \le$$

$$\le |D^2_{x_i} u_m|_{2,G}\, |D_{x_j} u_m|_{4,G}\, |D_{x_j} A_{i,m}|_{4,G} \le$$

$$\le K |D_{x_j} A_i|_{4,G}\, |D^2_{x_i} u_m|_{2,G} \left\{ |D^2_{x_j} u_m|^{1/2}_{2,G} \|u_m\|^{1/2}_{L^\infty(G)} + \|u_m\|_{L^\infty(G)} \right\}$$

$$\le K |D_{x_j} A_i|_{4,G}\, |D^2_{x_i} u_m|_{2,G} \left\{ |D^2_{x_j} u_m|^{1/2}_{2,G} \left( \frac{1}{\lambda} \sup_G |f| \right)^{1/2} + \left( \frac{1}{\lambda} \sup_G |f| \right) \right\}.$$

As a consequence of (4″) and (3) we can write:

$$(8) \qquad \forall m\ , \quad \|u_m\|_{W^{2,2}(G)} \le C \sup_G |f|$$

[using (4″), at the last member of (3) the same quantity of the first member appears with an exponent $< 1$]; from (8) and well known compactness results, at last one obtains:

**Theorem $\mathcal{M}_2$** — *In the hypotheses (2′), for all $f \in L^\infty(G)$, equation (1) has at least one solution $u \in W^{2,2}_{\#}(G)$.*

Let us notice that (8) gives neither an a priori estimate for $u \in W^{2,2}_{\#}(G)$ nor an unicity result; it is enough to remind the Alexandrov example (see also [GT] p.185), where the coefficients, for $n > 4$, are $W^{1,4}$.

On the other hand, the hypothesis: $f \in L^\infty(G)$ allows (via Maximum principle and Gagliardo - Nirenberg inequality) to bound $D_{x_j}u_m$ in $L^4(G)$, and therefore (see (4″)) to reduce (when $n > 4$) the summability of $D_{x_j}A_i$ from $L^n$ to $L^4$.

An other important result by C. Miranda ([M2], 1963) is an extension of Gagliardo - Nirenberg inequality, which we expose here in the simpler case: if $u \in L^{p_0}(G)$ and $|D_{x_i}u|^\alpha D^2_{x_i}u \in L^p(G)$ $(p \geq 1$ and $p_0 > 0)$, then with $\underline{\alpha \geq 0}$ and $q_1 \equiv \dfrac{pp_0(\alpha+2)}{p+p_0} \leq \eta(\alpha+2)$ one has:

$$(9) \qquad |D_{x_i}u|_{q_1,G} \leq K\left\{\left||D_{x_i}u|^\alpha D^2_{x_i}u\right|^{1/(\alpha+2)}_{p,G} |u|^{1/(\alpha+2)}_{p_0,G} + |u|_{p_0,G}\right\}$$

where $K$ depends only on $n$ and $\eta$.

Shortly afterwards L. Nirenberg [N2] (and in part C. Miranda) attained a different extension of Gagliardo - Nirenberg inequality: if $D^2_{x_i}u \in L^r(G)$ and $\underline{u \in C^{0,\beta}(G)}$ $[r \geq 1$ and $0 < \beta < 1]$, then the following inequality is valid:

$$(10) \qquad |D_{x_i}u|_{q,G} \leq K\left\{|D^2_{x_i}u|^a_{r,G}[u]^{1-a}_{\beta,G} + [u]_{\beta,G}\right\},$$

where: $a = \dfrac{1-\beta}{2-\beta}$, $q = \dfrac{2-\beta}{1-\beta}r$, $K$ depends only on $n$ and

$$[u]_{\beta,G} = \sup_{\substack{x',x''\in G\\ x'\neq x''}} \frac{|u(x') - u(x'')|}{|x'-x''|^\beta}.$$

At this point some remarks are suitable:

I) in the majorization of integral (4), the higher (stronger) the summability which we can assure to $D_{x_j}u$, the lower the summability which we are forced to suppose for the derivatives $D_{x_j}A_i$ of the coefficients; therefore, when $n > 4$, is preferable to use Gagliardo - Nirenberg instead of Sobolev inequality; even if this requires the renunciation to Unicity Theorem and the contracting to $L^\infty(G)$ of the class in which to assign the known term $f$;

II) in every case ($A_i \in W^{1,n}$, or else $A_i \in W^{1,4}$ and $n > 4$), the coefficients are *non* continuous; but an important difference exists between the two cases: if $A_i \in W^{1,n}$ then $A_i \in VMO$, while if $4 < n$ and $A_i \in W^{1,4}$ then it may be $A_i \notin VMO$; at this regard we recall that $VMO$ is the class of the locally summable functions $f$ for which:

$$\lim_{r \to 0} \sup_{\rho < r} \left\{ |B_\rho|^{-1} \int_{B_\rho} |f(x) - f_{B_\rho}| dx \right\} = 0 \tag{11}$$

where $B_\rho$ is the generic ball of radius $\rho$ and $f_{B_\rho} = |B_\rho|^{-1} \cdot \int_{B_\rho} f(x)dx$; (11) is equivalent to a sort of "average continuity";

III) on the other hand (whether $A_i \in W^{1,n}$ or $A_i \in W^{1,4}$ [and $n > 4$]) in any case the coefficients have summable distributional derivatives (with various exponents), which makes possible all partial integrations bringing to (3); therefore, thou C. Miranda genially uses non variational methods in dealing with (1), nevertheless this equation cannot be considered "drastically" non variational;

IV) after all we emphasize that Miranda method does not force to have recourse to limitative hypotheses (in a "quantitative" sense) as, for example, the Cordes condition:

$$\sum_{i,k=1}^{n} a_{ik}^2 \leq \frac{1}{n-1+\varepsilon} \left( \sum_{i=1}^{n} a_{ii} \right)^2 \quad ; \qquad 0 < \varepsilon < 1 . \tag{12}$$

## 2. A New Method

Remark III seems to prevent anybody from using Miranda technique when coefficients of the equation, besides discontinuous, are also lacking in summable distributional derivatives. However, at the beginning of '80's, I convinced myself that it is possible to overcome the obstacle, when the coefficients belong to suitable Sobolev - Slobodeczki Spaces $W^{s,p}$. Let us recall that, when $0 < s < 1$, $p \geq 1$ and $k$ integer $\geq 0$, $W^{k+s,p}(G)$ is the space of the functions $f \in W^{k,p}(G)$ such that:

$$\|f\|^p_{W^{k,p}(G)} + \sum_{|\alpha|=k} \iint_{G \times G} \frac{|D^\alpha f(x) - D^\alpha f(y)|^p}{|x-y|^{n+sp}} dx\, dy < +\infty . \tag{13}$$

One poses: $W^{k+s,p}_{\#}(G)$ = closure of $C^\infty_{\#}(\mathbf{R}^n)$ in $W^{k+s,p}(G)$. The important thing is that, posed $\Omega = G \times ]0,1[$, every $F \in W^{1,p}(\Omega)$ has a trace on $G$, and this trace is a function $f \equiv \gamma F \in W^{1-\frac{1}{p},p}(G)$; vice-versa every $f \in W^{1-\frac{1}{p},p}(G)$ is the trace on $G$ of a suitable $F \in W^{1,p}(\Omega)$. And more than that, there exists a linear continuous operator $\Re : W^{1-\frac{1}{p},p}(G) \to W^{1,p}(\Omega)$ such that:

$$\forall f \in W^{1-\frac{1}{p},p}(G) , \quad \gamma \Re f = f .$$

The operator $\Re$ is said a "relevament" (drawing), and $F \equiv \Re f$ is said a "relevament of $f$".

It is possible to define the operator $\Re$ in the following manner, for every $f \in W_{\#}^{1-\frac{1}{p},p}(G)$:

$$(14) \qquad (\Re f)(x,t) = \frac{1}{t^n} \int_{|y-x|\le t} \rho\left(\frac{y-x}{t}\right) f(y) dy , \qquad \begin{cases} x \in \mathbf{R}^n \\ t > 0 \end{cases}$$

where $\rho(z)$ is an averaging kernel:

$$(14') \quad \rho \in C_0^\infty(\mathbf{R}^n) ; \ \rho(z) \ge 0 , \quad \forall z; \quad \rho(z) = 0 \quad \text{for } |z| \ge 1; \quad \int_{|z|\le 1} \rho(z) dz = 1 .$$

It is easy to see that:

$$(14'') \qquad \forall f \in W_{\#}^{1-\frac{1}{p},p}(G) , \quad \Re f \in W_{\#}^{1,p}(\Omega) ,$$

where, posed $S = \mathbf{R}^n \times [0,1]$ and denoted $C_{\#}^\infty(S)$ the manifold of the $C^\infty(S)$-functions $2\pi$-periodic in $x_1, \dots, x_n$, we have defined the Spaces:

$$\begin{cases} C_{\#*}^\infty(S) = \{u \in C_{\#}^\infty(S) : D_t^2 u(x,0) = D_t^2 u(x,1) = 0\} , \\ W_{\#}^{k,p}(\Omega) = \text{closure of } \ C_{\#}^\infty(S) \ \text{ in } W^{k,p}(\Omega) , \\ W_{\#*}^{k,p}(\Omega) = \text{closure of } \ C_{\#*}^\infty(S) \ \text{ in } W^{k,p}(\Omega) . \end{cases}$$

It is also easy (see [CFL]) to verify that, if $0 < s < 1$ and $p \ge 1$:

$$sp = n \Longrightarrow W^{s,p}(G) \subset VMO ,$$

while, if $sp < n$, $W^{s,p}(G) \not\subset VMO$.

◇ Well, the idea is the following. We intend to solve the equation:

$$(1_\circ) \qquad -\sum_i^{1\dots n} a_i(x) D^2_{x_i} w + \lambda w = g(x) \ , \qquad g \in W^{\frac{1}{2},2}_{\#}(G)$$

with coefficients $a_i \in W^{\frac{n}{n+1},n+1}_{\#}(G)$ [and $\lambda > 0$]; by the relevaments $A_i \equiv \Re a_i$ of the coefficients $a_i$ and by the relevament $f \equiv \Re g$ of the known term $g$, we perform the auxiliary equation (*containing a "further variable"*):

$$(15) \qquad \begin{cases} -\sum_i^{1\dots n} A_i(x,t) D^2_{x_i} u - D^2_t u + \lambda u = f \ , & \text{in } \Omega \\ \\ u \in W^{3,2}_{\#*}(\Omega) \quad [\text{whence}: \ D^2_t u(x,0) = D^2_t u(x,1) = 0]; \end{cases}$$

in this equation, thanks to (14″), we have:

$$A_i \in W^{1,n+1}_{\#}(\Omega) \ , \quad f \in W^{1,2}_{\#}(\Omega) \quad (\Omega = G \times ]0,1[);$$

the problem (15) is a "mixed type" boundary problem rather unusual (periodicity in the variables $x_1, \dots, x_n$, and annihilating of the second derivatives respect to the variable $t$, for $t = 0$ and $t = 1$).

But the important thing is that the coefficients $A_i$ belong to $W^{1,n+1}_{\#}(\Omega)$; they possess consequently summable distributional derivatives (while the $a_i$ do not); so all partial integrations required by Miranda method are possible for problem (15).

If we find the way to solve the auxiliary problem (15) and if $u(x,t) \in W^{3,2}_{\#*}(\Omega)$ is one of its solutions, then $w(x) \equiv (\gamma u)(x) \equiv u(x,0) \equiv$ trace of $u$ on $G$ will be a solution of the problem $(1_\circ)$, because: $A_i(x,0) = a_i(x)$, $f(x,0) = g(x)$, $D^2_t u(x,0) = 0$. Naturally $w \equiv \gamma u \in W^{5/2,2}_{\#}(G)$.

In this way turns out outflanked the obstacle of non summability of the distributional derivatives of the coefficients $a_i(x)$; because we work on the coefficients $A_i(x,t)$ of Problem (15) which are of class $W^{1,n+1}$.

I consider interesting to have recourse to the trace spaces (as $W^{5/2,2}_{\#}(G)$) to seek the solutions of an equation (as $(1_\circ)$); while commonly they are used to assign the boundary values. Naturally, from this method we can not expect Unicity Theorems, but only Existence Theorems. Obviously the just now exposed idea is to be done concrete: the Miranda method is to be applied cleverly to equation (15), to obtain a solution $u \in W^{3,2}_{\#*}(\Omega)$.

Indeed we must *first* derive one time the equation (15) [initially written with $u \in C^\infty_{\#*}(S)$] (with respect to $x_k$ and $t$), and *last* multiply by suitable third derivatives $D^3 u(x,t)$ and integrate (by parts) on $\Omega$.

In this way we reach an inequality, analogous to (3), in which appear at the first member the (squared) seminorms:

$$|D^3u|^2_{2,\Omega} \quad \text{and} \quad \lambda|D^2u|^2_{2,\Omega} ,$$

and at the second [in place of the integrals of (4)] integrals of the type:

$$\int_\Omega D_{x_j}A_iD^2_{x_j}uD^3_{x_i}u \; . \tag{16}$$

Since $A_i \in W^{1,n+1}_{\#}(\Omega)$, the integrals (16) can be majorized by means of Sobolev Theorem, which gives for $D^2_{x_j}u$ exactly the summability exponent $2(n+1)/(n-1)$ when $D^3u \in L^2(\Omega)$. In this manner we reach at last an existence and unicity theorem for auxiliary problem (15) and an existence theorem for $(1_\circ)$.

A first result in this sense was obtained by myself and P. Zecca [CnZ] (1985), in the case $n = 3$; afterwards P. Zecca passed to the case of any $n$, considering also a more general equation than $(1_\circ)$ [Z1]; he also showed that for $n = 3$ a Unicity Theorem can be established (thanks to well known result by Pucci and Alexandrov, because $W^{1/2,2}(G) \subset L^{3\cdot 2/(3-\frac{1}{2}\cdot 2)}(G) = L^3(G) \equiv L^n(G)$, if $n = 3$).

But P. Zecca, always with the aim of reducing the regularity of the coefficients $a_i(x)$ of $(1_\circ)$, obtained in [Z2] also results analogous to Theorem $\mathcal{M}_2$ of C. Miranda, assuming: $a_i \in W^{5/6,6}_{\#}(G)$ [and consequently $A_i \in W^{1,6}_{\#}(\Omega)$], and assigning the known term $g$ of $(1_\circ)$ in $W^{\frac{1}{2},2}_{\#}(G) \cap L^\infty(G)$ [and consequently $f \equiv \Re g$ in $W^{1,2}_{\#}(\Omega) \cap L^\infty(\Omega)$]. Approximating in $W^{1,6}(\Omega)$ the $A_i$ by means of coefficients $A_{i,m}(x,t) \in C^\infty_{\#}(S)$, and approximating $f$ in $W^{1,2}(\Omega)$ by means of known terms $f_m(x,t) \in C^\infty_{\#}(S) \cap L^\infty(\Omega)$ [with $\sup_\Omega |f_m| \le \sup_\Omega |f|, \;\; \forall m$], P. Zecca considers in [Z2] the "auxiliary approximated" problem::

$$\begin{cases} -\sum\limits_i^{1\ldots n} A_{i,m}(x,t)D^2_{x_i}u - D^2_t u + \lambda u = f_m(x,t), \quad \text{in } \; \Omega \\[2em] u \equiv u(x_1,\ldots,x_n,t) \quad \text{periodic in } \; x_1,\ldots,x_n \\[2em] D^2_t u(x,0) = D^2_t u(x,1) = 0 \; . \end{cases} \tag{$15_m$}$$

The approximating problem $(15_m)$ has one and only one solution $u_m(x,t) \in C^\infty_{\#*}(S)$, for which - thanks to maximum principle - the following estimate is valid:

$$\sup_\Omega |u_m| \le \frac{1}{\lambda}\sup_\Omega |f_m| \le \frac{1}{\lambda}\sup_\Omega |f| \, , \quad \forall m \; . \tag{17}$$

Deriving (one time) ($15_m$) [written for $u_m$], multiplying by suitable third derivatives of $u_m$ and then integrating by parts on $\Omega$, one obtains an inequality [which we refer to ($3_m$)] analogous to (3), where, at the first member, the (squared) seminorms appear:

$$|D^3 u_m|^2_{2,\Omega} \quad \text{and} \quad \lambda|D^2 u_m|^2_{2,\Omega} \ ,$$

and, at the second member, in place of integrals (4), some integrals of type (16) with $u_m$ in place of $u$. For such integrals, thanks to (17) and to Gagliardo - Nirenberg inequality, one has:

$$(16_m) \qquad |\int_\Omega D_{x_j} A_{i,m} D^2_{x_j} u_m D^3_{x_i} u_m| \leq$$

$$\leq |D^3_{x_i} u_m|_{2,\Omega} |D_{x_j} A_{i,m}|_{6,\Omega} |D^2_{x_j} u_m|_{3,\Omega} \leq$$

$$\leq K|D_{x_j} A_i|_{6,\Omega} |D^3_{x_i} u_m|_{2,\Omega} \left\{ |D^3_{x_j} u_m|^{2/3}_{2,\Omega} \left( \sup_\Omega |u_m| \right)^{1/3} + \left( \sup_\Omega |u_m| \right) \right\} \leq$$

$$\leq K|D_{x_j} A_i|_{6,\Omega} |D^3_{x_i} u_m|_{2,\Omega} \left\{ |D^3_{x_j} u_m|^{2/3}_{2,\Omega} \left( \frac{1}{\lambda} \sup_\Omega |f| \right)^{1/3} + \left( \frac{1}{\lambda} \sup_\Omega |f| \right) \right\} .$$

From ($16_m$) and ($3_m$) it follows:

$$(18) \qquad \forall m \ , \quad \|u_m\|_{W^{3,2}(\Omega)} \leq C \sup_\Omega |f| \ ,$$

from which, thanks to well known compactness theorems, one obtains the existence of at least one solution $u(x,t) \in W^{3,2}_{\#*}(\Omega)$ of the problem (15).The trace $w(x) \equiv (\gamma u)(x) \equiv u(x,0)$ will be a solution $\in W^{5/2,2}_{\#}(G)$ of the problem ($1_\circ$) [with $g \in W^{1/2,2}_{\#}(G) \cap L^\infty(G)$].

Let us remark that, when $n+1 > 6$, the hypothesis: $A_i \in W^{1,6}_{\#}(\Omega)$ (i.e.: $a_i \in W^{5/6,6}_{\#}(G)$) [necessary to write ($16_m$)] is better than hypothesis: $a_i \in W^{\frac{n}{n+1},n+1}_{\#}(G)$. When $\underline{n=5}$ the two hypotheses coincide. In both cases ($a_i \in W^{\frac{n}{n+1},n+1}_{\#}(G)$ or else $a_i \in W^{5/6,6}_{\#}(G)$ with $n > 5$) the coefficients are discontinuous and lacking in summable distributional derivatives; but - as already said - an important difference exists between the two cases: if $a_i \in W^{\frac{n}{n+1},n+1}_{\#}(G)$, then $a_i \in VMO$; if $a_i \in W^{5/6,6}_{\#}(G)$ and $n > 5$, then it is possible that $a_i \notin VMO$; obviously if $a_i \in W^{5/6,6}_{\#}(G)$ and $n = 5$, then $a_i \in VMO$.

## 3. Some Important Results

In two important papers [CFL1] and [CFL2]. F. Chiarenza, M. Frasca and P. Longo study the elliptic equations with $VMO$ coefficients.

As we know, in this class are included the Miranda equations with $W^{1,n}$, as well the equations of [CnZ] and [Z1] with $W_{\#}^{\frac{n}{n+1},n+1}$ coefficients. On the contrary, it is impossible to make use of the $VMO$ Theory in dealing with equations with $W^{1,4}$ coefficients (if $n > 4$), as well with equations with $W_{\#}^{5/6,6}$ (if $n > 5$). $VMO$ Theory turns again applicable in the limit cases: $W^{1,4}$ coefficients and $n = 4$, $W_{\#}^{5/6,6}$ coefficients and $n = 5$.

Well, in a recent paper [Cn2] I study some elliptic equations with (bounded periodic) $W^{1-1/p,p}(G)$ coefficients, where $n \geq 5$ and $p$ is "a little lower than 6"; therefore also if $n = 5$ one has: $sp = \left(1 - \frac{1}{p}\right) p < 5 = n$ and $VMO$ Theory is not applicable

Being, consequently, $p < 6$ the summability exponent of $DA_i$ [remember that $A_i \equiv \Re a_i$], a suitable summability has been necessary to assure to the second derivatives $D_{x_i}^2 u$ (see integrals (16) and ($16_m$)). To this end, we establish an integral inequality which is the blending of the Miranda inequality (9) [but with $\alpha < 0$] with the Nirenberg one (10).

Precisely, posed:

$$(19)\quad \begin{cases} r = 2 + 2\alpha + \dfrac{2}{2-\beta}\,, \quad a = \dfrac{1-\beta}{2-\beta}\,, \quad \text{with}: -\dfrac{1}{2} \leq \alpha < 0 < \beta \leq \dfrac{1}{2} \\[2ex] D_2^3 u = \displaystyle\sum_{i,j}^{1\ldots n} (D_{x_j}^2 D_{x_i} u)^2 + \sum_{i,j}^{1\ldots n} (D_{x_j} D_{x_i} D_t u)^2 + \sum_{i}^{1\ldots n} (D_{x_i} D_t^2 u)^2 + (D_t^3 u)^2 \\[2ex] \theta = \left[1 + \displaystyle\sum_{j}^{1\ldots n} (D_{x_j}^2 u)^2 + \sum_{j}^{1\ldots n} (D_{x_j} D_t u)^2 + (D_t^2 u)^2\right]^{1/2} \equiv [1 + D_2^2 u]^{1/2}\,, \end{cases}$$

we prove that:

$$(20)\ \forall\, u \in C_{\#*}^{\infty}(S)\,, \quad \left(\int_{\Omega} |\theta|^r\right)^{1/r} \leq$$

$$\leq C \left\{ \left(\int_{\Omega} \theta^{2\alpha} D_2^3 u\right)^{1/(2\alpha+4-2a)} [u]_{\beta,\Omega}^{(1-a)/(\alpha+2-a)} + \left(\int_{\Omega} \theta^{2\alpha} D_2^3 u\right)^{1/(2\alpha+4)} [u]_{\beta,\Omega}^{1/(\alpha+2)} + 1 \right\}$$

with $C$ depending only on $n$.

Moreover we make use of a fundamental result of N.V. Krylov and M.V. Safonov [KS1] [KS2] (see also [GT] Corollaries 9.24, 9.28 and 9.29) concerning the Hölder continuity of the solutions of the *non variational* elliptic equations; this result extends to the non variational case the famous De Giorgi Theorem of 1957 [DG]. Krylov and Safonov individuate, for the solutions of the equations with known term $\in L^n$, an Hölder exponent which substantially depends on $n$ and on maximal and minimal eigenvalues of the coefficients matrix.

Let us go into some detail of [Cn2]. We consider, of course, the equation:

$$(1_\circ) \qquad \begin{cases} -\sum_i^{1\dots n} a_i(x) D^2_{x_i} w + \lambda w = g(x) \;, & \text{in } \; G \equiv ]0,2\pi[^n \;, \\ \\ g \;, \; w \qquad 2\pi - \text{periodic in } \; x_1, \dots, x_n \;. \end{cases}$$

The hypotheses are:

$$(21) \qquad \begin{cases} \lambda > 0 \;, \qquad 1 \le a_i(x) \le \Lambda \;, \\ \\ a_i \in W_{\#}^{1-\frac{1}{p},p}(G) \quad \text{with :} \quad p = 2 + 2(1+\alpha)(2-\beta), \\ \\ \text{where :} \;\; -[2\Lambda(24n^2 + 28n + 4)]^{-1} \le \alpha < 0 \\ \\ \text{while } \beta \equiv \beta(n,\Lambda) \;\; \text{is the Hölder exponent of} \\ \\ \text{Krylov and Safonov for the elliptic operators} \\ \\ \text{whose coefficients matrix has 1 and } \Lambda \text{ as} \\ \\ \text{minimal and maximal eigenvalues.} \end{cases}$$

So $p \equiv 2 + 2(1+\alpha)(2-\beta) < 6$ depends on $\alpha, \beta$, that is on $\Lambda$.

We establish the following

**Existence Theorem** — *In the hypotheses (21), for every* $g \in W_{\#}^{1/2,2}(G) \cap L^\infty(G)$

*equation* $(1_\circ)$ *has at least one solution* $w \in W_{\#}^{3-1/s,s}(G)$, *where:* $s{=}2+2\dfrac{2-\beta}{3-\beta}\alpha \in ]1,2[$.

The general plan of the proof is analogous to [CnZ] one. We perform the functions $A_i \equiv \Re a_i$ and $f \equiv \Re g$ having respectively as traces for $t=0$ the coefficients $a_i$ and the known term $g$ of $(1_\circ)$.

$A_i$ and $f$ verify in $\Omega = G\times]0,1[$:

$$(22)\qquad \begin{cases} A_i \in W_{\#}^{1,p}(\Omega)\ , \quad 1 \le A_i(x,t) \le \Lambda \\ \\ f \in W_{\#}^{1,2}(\Omega) \cap L^\infty(\Omega)\ , \quad \sup\limits_{\Omega}|f| \le \sup\limits_{G}|g|\ . \end{cases}$$

Then we approximate $A_i$ in $W^{1,p}(\Omega)$ by means of a sequence $A_{i,m} \in C_{\#}^\infty(S)$ and $f$ in $W^{1,2}(\Omega)$ by means of a sequence $f_m \in C_{\#}^\infty(S)$ [such that: $\forall m\ ,\ \sup\limits_{\Omega}|f_m| \le \sup\limits_{\Omega}|f| \le \sup\limits_{\Omega}|g|$]. Then we solve the problems:

$$(23_m)\qquad \begin{cases} \left[-\sum\limits_i^{1\dots n} A_{i,m}(x,t)D_{x_i}^2 - D_t^2 + \lambda\right] u(x,t) = f_m(x,t)\ , \quad \text{in } \Omega \\ \\ u \in C_{\#*}^\infty(S) \end{cases}$$

At last for the solutions $u_m(x,t)$ of $(23_m)$ we write majorizations so "good" that - thanks to known compactness theorems (Rellich - Kondrachov) - from the traces sequences $u_m(x,0)$ we can extract a subsequence converging to a function $w \in W_{\#}^{3-1/s,s}(G)$ which solves equation $(1_\circ)$.

The "good" majorizations for the $u_m$ are obtained deriving $(23_m)$, multiplying by suitable factors of the type:

$$\theta^{2\alpha}D^3u_m \equiv \left[1+\sum_j^{1\dots n}(D_{x_j}^2u_m)^2+\sum_j^{1\dots n}(D_{x_j}D_tu_m)^2+(D_t^2u_m)^2\right]^\alpha D^3u_m$$

and integrating by parts on $\Omega$.

In this manner we reach an estimate [analogous to (3)], in which, at a first member, the integrals appear:

$$(24)\qquad \int_\Omega \theta^{2\alpha}D_2^3u_m + \lambda\int_\Omega \theta^{2\alpha}D_2^2u_m\ ,$$

while, at the second member, in place of $(16_m)$, appear integrals of the type :

$$(25)\qquad \int_\Omega D_{x_j}A_{i,m}\cdot D_{x_j}^2u_m\cdot\theta^{2\alpha}\cdot D_{x_i}^3u_m \equiv I_m\ .$$

Being: $|D^2_{x_j}u_m| \le \theta$, $|D^3_{x_i}u_m|^2 \le D^3_2 u_m$, and also (see (19) and (21)): $\frac{2p(1+\alpha)}{p-2} = = r \equiv 2 + 2\alpha + \frac{2}{2-\beta}$, the integrals (25) are bounded as follows:

$$(26) \qquad \begin{aligned} |I_m| \;&\le\; \int_\Omega |D_{x_j}A_{i,m}| \cdot \theta^\alpha |D^3_{x_i}u_m| \cdot \theta^\alpha |D^2_{x_j}u_m| \le \\ &\le\; \left(\int_\Omega |D_{x_j}A_{i,m}|^p\right)^{1/p} \left(\int_\Omega \theta^{2\alpha} D^3_2 u_m\right)^{1/2} \left(\int_\Omega \theta^{\frac{2p(1+\alpha)}{p-2}}\right)^{\frac{p-2}{2p}} \le \\ &\le\; K \left(\int_\Omega |D_{x_j}A_i|^p\right)^{1/p} \left(\int_\Omega \theta^{2\alpha} D^3_2 u_m\right)^{1/2} \left[\left(\int_\Omega \theta^r\right)^{1/r}\right]^{1+\alpha} . \end{aligned}$$

And here we have recourse to the (ugly but good) inequality (20) and to Krylov - Safonov Theorem to bound the last factor of the last member of (26). Let us remark that to lower $p$ below 6 it is necessary to raise $\frac{2p}{p-2}$ above 3; but the effect on the summability exponent of the second derivatives $D^2u_m$ (i.e. of $\theta$) is reduced by the factor $(1+\alpha)$ with $\underline{\alpha < 0}$: here is manifest the flexibility of inequality (20). So much that, by (20), the last factor of the last member of (26) turns out to be bounded by the first of the integrals (24) raised to suitable exponents. We attain so an inequality whose first member is given by (24), and whose second member has the form:

$$(24') \qquad \sum_j^{1\dots k} C_j \left(\int_\Omega \theta^{2\alpha} D^3_2 u_m\right)^{\alpha_j} , \quad \text{with}: \; 0 \le \alpha_1, \dots, \alpha_k < 1$$

At this point, thanks to known numerical inequalities, we reach a majorization of the type:

$$(27) \qquad \forall m , \quad \int_\Omega \theta^{2\alpha} D^3_2 u_m \le K_0 \;\; \text{independent on } m .$$

Since the first member of (27) is substantially bounded from below by the norm $\|u_m\|_{W^{3,s}(\Omega)}$, (27) is the "good" majorization we need. The Theorem is proved.

In a former paper [Cn1] (1989) I had proved [having recourse to an inequality similar to (20)] an other Existence Theorem for elliptic equations with $W^{1,4+2\alpha}$ coefficients in $]0,1[^n$ where: $n \ge 4$, and $-\frac{1}{6} \le \alpha < 0$. For example, for the equation:

$$(28) \qquad \sum_i^{1\dots 4} [2 + \sin(|x - x^i|^{-2\beta_i})] D^2_{x_i} u = f(x)$$

where: $x^i \equiv (x^i_1, \dots, x^i_4) \in ]0,1[^4$ , $\quad 0 < \beta_i < \frac{1}{22}$ .

Equation (28) cannot be dealt neither with $VMO$ Theory [because the coefficients belong to $W^{1,4+2\alpha}$ and $4+2\alpha < 4 \equiv n$], nor with Miranda Theorem $\mathcal{M}_2$ [because (28) does not fulfill (2′)].

At this point, in this "defiance" to go beyond the class $VMO$ for the coefficients of elliptic equations, we please to present an equation which can be faced only with the help of the existence theorem of [Cn2]. But to this end some bustle is required.

## 4. A Digression on the Cantor Set

Let us begin with a formal construction of the open set $A \subset [0,1]$ which is the complement in $[0,1]$ of the famous "Cantor set".
Fixed $k$ positive integer, let us divide $[0,1]$ into the intervals:

$$I_1 \equiv \left]\frac{1}{2} - \frac{1}{2}\left(\frac{1}{2k+1}\right) \ , \ \ \frac{1}{2} + \frac{1}{2}\left(\frac{1}{2k+1}\right)\right[$$

(of length $\frac{1}{2k+1}$), $I_1^0 = \left[0, \frac{1}{2} - \frac{1}{2}\left(\frac{1}{2k+1}\right)\right]$ ("left wing" of length $\frac{k}{2k+1}$) and

$$I_1^1 \equiv \left[\frac{1}{2} + \frac{1}{2}\left(\frac{1}{2k+1}\right), 1\right]$$

("right wing" of length $\frac{k}{2k+1}$).

We call $I_1$: "0-generation" interval. Proceeding by induction, we suppose, for $r$ integer $\geq 1$, already defined the $2^{r-1}$ "$r-1$-generation intervals":

$$(29)\qquad \begin{cases} I_{j_1 \ldots j_{r-1} j_r} \equiv I_{j_1 \ldots j_{r-1} 1} \ , \quad \text{open interval with :} \\ \\ j_r = 1; \ \ j_l \in \{0,1\} \text{ if } l \leq r-1 \ , \quad |I_{j_1 \ldots j_r}| = \left(\dfrac{k}{2k+1}\right)^{r-1} \left(\dfrac{1}{2k+1}\right) \ ; \end{cases}$$

[for $r = 1$, we already know $I_1$].
For the generic $I_{j_1 \ldots j_r}$ of the $r-1$-generation, let us indicate by $I^0_{j_1 \ldots j_{r-1} 1}$ its "left wing" (i.e. the closed interval adjacent to $I_{j_1 \ldots j_r}$ on the left) of length $\left(\frac{k}{2k+1}\right)^r$; and by $I^1_{j_1 \ldots j_{r-1} 1}$ its "right wing" of length $\left(\frac{k}{2k+1}\right)^r$. Let us indicate, at last, by $I_{j_1 \ldots j_{r-1} 01}$ "the central part" of (i.e. the open interval concentric to) $I^0_{j_1 \ldots j_{r-1} 1}$ of length $\left(\frac{k}{2k+1}\right)^r \left(\frac{1}{2k+1}\right)$, and by $I_{j_1 \ldots j_{r-1} 11}$ "the central part" of $I^1_{j_1 \ldots j_{r-1} 1}$ of length

$\left(\dfrac{k}{2k+1}\right)^r\left(\dfrac{1}{2k+1}\right)$. Remain so defined all $2^r$ "$r$-generation intervals": $I_{j_1\ldots j_r j_{r+1}}$ (fulfilling (29) with $r+1$ in place of $r$). Springs in this manner a numerable family $\{I_{j_1\ldots j_{r-1}j_r}\}$ which can be ordered in a "genealogical tree":

generation 0: $I_1$ ($\swarrow$ $\searrow$)

generation 1: $I_{01}$ ($\swarrow$ $\searrow$) $I_{11}$ ($\swarrow$ $\searrow$)

generation 2: $I_{001}$ $I_{011}$ $I_{101}$ $I_{111}$

/ \ / \ / \ / \

The wings [of length $\left(\dfrac{k}{2k+1}\right)^r$] of the $(r-1)$-generation intervals will be called "$(r-1)$-generation wings". If $I_{j_1\ldots j_r 1}$ is the generic interval of generation $r \geq 1$, by construction we have:

$$\begin{cases} I_{j_1\ldots j_r 1} \text{ is open and concentric to } I^{j_r}_{j_1\ldots j_{r-1}1} \quad [j_r = 0,1] \\[2ex] |I^0_{j_1\ldots j_r 1}| + |I_{j_1\ldots j_r 1}| + |I^1_{j_1\ldots j_r 1}| = 2\left(\frac{k}{2k+1}\right)^{r+1} + \left(\frac{k}{2k+1}\right)^r\left(\frac{1}{2k+1}\right) = \left(\frac{k}{2k+1}\right)^r = |I^{j_r}_{j_1\ldots j_{r-1}1}| \end{cases}$$

and:

$$I^0_{j_1\ldots j_r 1} \cup I_{j_1\ldots j_r 1} \cup I^1_{j_1\ldots j_r 1} = I^{j_r}_{j_1\ldots j_{r-1}1}\,. \tag{29'}$$

By induction on $r$, one easily checks that:

(29″) every $(r-1)$-generation wing is disjointed from *all* the intervals of generation $0, 1, \ldots, r-1$ and from *all* the remaining $(r-1)$-generation wings.

As a corollary we have:

(29‴) every $r$-generation interval is disjointed from *all* the intervals of generation $0, 1, \ldots, r-1$ and from all the remaining $r$-generation intervals.

As a consequence we have

**Proposition 1** — The open $A \equiv \bigcup_{\substack{r \text{ integer } \geq 1 \\ j_l \in \{0,1\} \text{ if } l \leq r-1 \\ j_r = 1}} I_{j_1 \ldots j_{r-1} j_r}$ has measure $= 1$.

**Proof.** In fact [by (29)] we have:

$$(30) \qquad |A| = \sum_{j_1, \ldots, j_{r-1}=0,1}^{r \geq 1; j_r = 1} |I_{j_1 \ldots j_r}| = \sum_{r=1}^{\infty} 2^{r-1} \left(\frac{k}{2k+1}\right)^{r-1} \left(\frac{1}{2k+1}\right)$$
$$= \frac{1}{2k+1} \sum_{r-1=0}^{\infty} \left(\frac{2k}{2k+1}\right)^{r-1} = \frac{1}{2k+1} \left(\frac{1}{1 - \frac{2k}{2k+1}}\right) = 1 \,,$$

having taken into account that [by (29‴)] the intervals $I_{j_1 \ldots j_r}$ are two by two disjointed. q.e.d.

Therefore the open set $A \subset [0,1]$ differs from $[0,1]$ by a set of measure zero (the Cantor set). A simple property of $A$ [which is proved by induction on $r$] is the following:

$$(31) \quad \begin{cases} \text{called } \sigma : \mathbf{R} \to \mathbf{R} \text{ the symmetry with center } 1/2 \quad (\sigma(x) = 1 - x) \,, \\[2ex] \text{for all } I_{j_1 \ldots j_r} \text{ we have:} \\[2ex] \sigma I_{j_1 \ldots j_{r-1} 1} = I_{j'_1 \ldots j'_{r-1} 1} \text{ where: } j'_l = \begin{cases} 1, & \text{if } j_l = 0 \\ 0, & \text{if } j_l = 1 \end{cases} \quad (l \leq r-1) \,. \end{cases}$$

An other important property of $A$ is expressed by the

**Proposition 2** — If $s > r \geq 1$, then the interval $I_{j_1 \ldots j_{r-1} \overbrace{10\ldots01}^{s-r}} \equiv I_{j_1 \ldots j_r 0 \ldots 0 j_s}$ lies to the right of $I_{j_1 \ldots j_r}$ at a distance $\left(\frac{k}{2k+1}\right)^s$; in other words:

$$(32) I_{j_1 \ldots j_r 0 \ldots 0 j_s} \equiv$$

$$\equiv ] \sup I_{j_1 \ldots j_r} + \left(\frac{k}{2k+1}\right)^s , \quad \sup I_{j_1 \ldots j_r} + \left(\frac{k}{2k+1}\right)^s + \left(\frac{k}{2k+1}\right)^{s-1} \left(\frac{1}{2k+1}\right) [ .$$

Proposition 2 is proved by induction on $q = s - r$. As an immediate corollary, we get:

$$(32') \qquad \{s > r \geq 1 ; \quad x \in I_{j_1 \ldots j_r} ; \quad y \in I_{j_1 \ldots j_r 0 \ldots 0 j_s}\}$$

$$\Longrightarrow |y - x| \geq \left(\frac{k}{2k+1}\right)^s .$$

An other corollary of (32) and (31) is the

**Proposition 3** — If $s > r \geq 1$, then the interval $I_{j_1 \ldots j_{r-1} \overbrace{01\ldots11}^{s-r}} \equiv I_{j_1 \ldots j_{r-1} 0 j_{r+1} \ldots j_s}$ lies to the left of $I_{j_1 \ldots j_{r-1} 1}$ at a distance $\left(\frac{k}{2k+1}\right)^s$. In general $I_{j_1 \ldots j_{r-1} 0 j_{r+1} \ldots j_{s-1} 1}$ lies to the left of $I_{j_1 \ldots j_{r-1} 1}$.

Ultimately we have:

**Lemma 1** — In order that the interval $I' \equiv I_{j'_1 \ldots j'_{s'}}$ lies to the left of $I'' \equiv I_{j''_1 \ldots j''_{s''}}$ it is necessary and sufficient that one of the following conditions (a), (b) be fulfilled:

(a) $\quad s' < s'' ; \quad j'_1 = j''_1 \equiv j_1 , \ldots , j'_{s'} = j''_{s'} \equiv j_{s'} = 1 ; \quad I' = I_{j_1 \ldots j_{s'}} , \quad I'' = I_{j_1 \ldots j_{s'} j''_{s'+1} \ldots j''_{s''}}$ ;

(b) $1 \leq r < s' ; \quad j'_r = 0 < j''_r = 1 ; \quad j'_1 = j''_1 \equiv j_1 , \ldots , j'_{r-1} = j''_{r-1} \equiv j_{r-1} \quad (\text{if } r > 1) ;$

$$I' = I_{j_1 \ldots j_{r-1} 0 j'_{r+1} \ldots j'_{s'}} ,$$

$$I'' = \begin{cases} I_{j_1 \ldots j_{r-1} 1} \quad (\text{if } s'' = r) \\ \\ I_{j_1 \ldots j_{r-1} 1 j''_{r+1} \ldots j''_{s''}} \quad (\text{if } s'' > r) \end{cases}$$

At this point, let us define in $A = \cup I_{j_1 \ldots j_r}$ the "Cantor function" (the "devil scale"):

$$c(x) = \frac{j_1}{2} + \frac{j_2}{2^2} + \ldots + \frac{j_r}{2^r}\,, \quad \text{for } x \in I_{j_1 \ldots j_r}\,; \tag{33}$$

obviously $\forall x \in A$, $0 < c(x) < 1$; and moreover:

$$c'(x) = 0\,, \quad \forall x \in A \{\Longrightarrow c'(x) = 0 \quad \text{a.e. in } [0,1]\}\,. \tag{33'}$$

Consequently we have

**Proposition 4** — The function $c(x)$ is *not* absolutely continuous (and hence not Lipschitz continuous).

Let now:

$$(34) x \in I_{j_1 \ldots j_r} \equiv I_{J_1 \ldots J_{r-1} 1}\,, \quad y \in I_{j_1 \ldots j_r 0 \ldots 0 j_{r_1} 0 \ldots 0 j_{r_2} 0 \ldots\ldots\ldots 0 j_s} \equiv I_{j_1 \ldots j_{r-1} 10 \ldots 010 \ldots 010 \ldots\ldots\ldots 01}$$

where: $r_0 \equiv r$, and $(\forall i \geq 0)$ $r_{i+1}$ is the index $m$ with $j_m = 1$ immediately following $r_i$; in other words:

$$\begin{cases} r \equiv r_0 < r_1 < r_2 < \ldots < r_q \equiv s\,; \quad j_r = j_{r_1} = j_{r_2} = \ldots = j_s = 1\,, \\ \\ j_l = 0 \quad \text{for } r < l \notin \{r_1, \ldots, r_q\}. \end{cases}$$

In particular: $r_1$ is the index $m$ with $j_m = 1$ immediately following $r$.

From (34) and (33) we get:

$$|c(x) - c(y)| = \frac{1}{2^{r_1}} + \frac{1}{2^{r_2}} + \ldots + \frac{1}{2^s}\,. \tag{35}$$

From (34), (32′) and Proposition 2 we get also:

$$|x - y| \geq \left(\frac{k}{2k+1}\right)^{r_1} + \left(\frac{k}{2k+1}\right)^{r_2} + \ldots + \left(\frac{k}{2k+1}\right)^{s}\,. \tag{36}$$

**Proposition 5** — Let: $0 < \lambda < 1$. If the condition (34) is fulfilled, then:

$$\left(1 - \frac{1}{r_1}\right)(\log 2)/\log\left(2 + \frac{1}{k}\right) \geq \lambda \Longrightarrow |c(x) - c(y)| \leq |x - y|^{\lambda}\,; \tag{37}$$

and hence a fortiori (because $r_1 > r$):

$$(37') \qquad \left(1-\frac{1}{r}\right)(\log 2)/\log\left(2+\frac{1}{k}\right) \geq \lambda \Longrightarrow |c(x)-c(y)| \leq |x-y|^\lambda .$$

**Proof.** — Thanks to (35) and (36) the following implications are valid:

$$\begin{aligned}
&|c(x)-c(y)| \leq |x-y|^\lambda \Longleftarrow \\
&\Longleftarrow \frac{1}{2^{r_1}}+\ldots+\frac{1}{2^s} \leq \left[\left(\frac{k}{2k+1}\right)^{r_1}+\ldots+\left(\frac{k}{2k+1}\right)^s\right]^\lambda \\
&\Longleftrightarrow \log\left\{\frac{1}{2^{r_1}}\left(1+\frac{1}{2^{r_2-r_1}}+\ldots+\frac{1}{2^{s-r_1}}\right)\right\} \leq \\
&\quad \leq \lambda \log\left\{\frac{1}{(2+\frac{1}{k})^{r_1}}\left[1+\frac{1}{(2+\frac{1}{k})^{r_2-r_1}}+\ldots+\frac{1}{(2+\frac{1}{k})^{s-r_1}}\right]\right\} \\
&\Longleftrightarrow r_1\log 2-\log\left(1+\frac{1}{2^{r_2-r_1}}+\ldots+\frac{1}{2^{s-r_1}}\right) \geq \\
&\quad \geq \lambda\left\{r_1\log\left(2+\frac{1}{k}\right)-\log\left[1+\left(2+\frac{1}{k}\right)^{r_1-r_2}+\ldots+\left(2+\frac{1}{k}\right)^{r_1-s}\right]\right\} \\
&\Longleftarrow r_1\log 2-\log 2 \geq \lambda r_1 \log\left(2+\frac{1}{k}\right) .
\end{aligned}$$

Implication (37) is thus established; and hence (37′) too. q.e.d..

Well, in correspondence with the assigned $\lambda \in ]0,1[$ *let us fix* $k \in \mathbf{N}$ such that:

$$\lambda < (\log 2)/\log\left(2+\frac{1}{k}\right) ;$$

afterwards let us fix $p \equiv p(\lambda) \in \mathbf{N}$ such that $p \geq 2$, and:

$$(38) \qquad \left(1-\frac{1}{p-1}\right)\frac{(\log 2)}{\log(2+\frac{1}{k})} < \lambda \leq \left(1-\frac{1}{p}\right)(\log 2)/\log\left(2+\frac{1}{k}\right) .$$

Let us consider all the pairs of intervals:

$$(39) \qquad I_{j_1\ldots j_r}\,, \quad I_{j_1\ldots j_r\ldots j_s} \quad \text{with}: \ 1 \leq r < s \leq p ;$$

thanks to Proposition 2, $\mathrm{dist}\{I_{j_1\ldots j_r}, I_{j_1\ldots j_r\ldots j_s}\} > 0$ for all these pairs; being these pairs in finite number, trivially we have:

$$(40) \qquad 0 < \delta \equiv \min_{1\leq r<s\leq p} \mathrm{dist}\{I_{j_1\ldots j_r},\ I_{j_1\ldots j_r\ldots j_s}\} ;$$

obviously $\delta$ depends on $\lambda$: $\delta = \delta(p) = \delta(p(\lambda))$. Now, if $x$ and $y$ fulfil (34) [i.e.: $x \in I_{j_1\ldots j_r}\,,\ \ y \in I_{j_1\ldots j_r\ldots j_s}$] with $1 \leq r < s \leq p$, trivially we have:

$$(41) \qquad |c(x)-c(y)| \leq 2 \equiv \left(\frac{2}{\delta^\lambda}\right)\delta^\lambda \leq \left(\frac{2}{\delta^\lambda}\right)|x-y|^\lambda .$$

If $x$ and $y$ fulfil (34) with $p \leq r < s$, then (38) and (37′) give:

$$(41') \qquad |c(x) - c(y)| \leq |x - y|^\lambda \; .$$

It remains to examine the case that $x$ and $y$ fulfil (34) with: $1 \leq r < p < s$. If all the indices $r_1 < r_2 < \ldots r_q \equiv s$ of (34) are $> p$, then in particular $r_1 > p$ and hence (38) (37) immediately give:

i) $\qquad |c(x) - c(y)| \leq |x - y|^\lambda \; .$

On the contrary, if $r_1 \leq p$, let $s'$ the greatest of the indices $r_1, \ldots, r_l, \ldots, r_q \equiv s$ of (34) such that $r_l \leq p$: whence:

ii) $\qquad 1 \leq r < s' \leq p < s'' \leq s$

where $s''$ is the index $r_l$ immediately following $s'$ [therefore (see (34)) $s''$ is the index $m$ with $j_m = 1$ immediately following $s'$]. Fixed a point $z \in I_{j_1 \ldots j_r \ldots j_{s'}}$ thanks to (ii) and (41) [and (32)] we get:

iii) $\qquad |c(z) - c(x)| \leq \left(\frac{2}{\delta^\lambda}\right) |z - x|^\lambda < \left(\frac{2}{\delta^\lambda}\right) |x - y|^\lambda \; ;$

while [observing that for the points $z \in I_{j_1 \ldots j_r \ldots j_{s'}}$ and $y \in I_{j_1 \ldots j_{s'} \ldots j_s}$ the index $s''$ accomplishes the same task as $r_1$ (see (34)) for the points $x$ and $y$] thanks to (ii), (38), (37) and (32) we get:

iv) $\qquad |c(y) - c(z)| \leq |y - z|^\lambda \leq |x - y|^\lambda .$

From (i), (iii) and (iv) it follows, in every case:

$$(41'') \qquad \begin{cases} \text{if } x \text{ and } y \text{ fulfil (34) with } 1 \leq r < p < s, \text{ then} \\ |c(x) - c(y)| \leq \left(1 + \left(\frac{2}{\delta^\lambda}\right)\right) |x - y|^\lambda \; . \end{cases}$$

From (41), (41′) and (41″) we get:

$$(42) \qquad \left. \begin{cases} x \text{ and } y \text{ fulfil (34); i.e.} \\ x \in I_{j_1 \ldots j_r} \text{ and } y \in I_{j_1 \ldots j_r \ldots j_s} \end{cases} \right\} \Rightarrow |c(x) - c(y)| \leq \left(\frac{2}{\delta^\lambda} + 1\right) |x - y|^\lambda \; ,$$

where $\delta \equiv \delta(p) \equiv \delta(p(\lambda))$ is given by (40).

An important property of $c(x)$ is expressed by

**Proposition 6** — The graph of the function $c(x)$ is symmetric, in the plane, with respect to the point $(1/2, 1/2)$. Precisely [taking in mind (see (31)) that $\sigma(x) \equiv 1 - x$ is the symmetry, in $\mathbf{R}$, of center $1/2$] we get:

$$\forall x \in A\,, \quad (\sigma(x),\ \sigma(c(x))) = (\sigma(x), c(\sigma(x)))\,, \tag{43}$$

i.e.:

$$\forall x \in A\,, \quad c(\sigma(x)) = \sigma(c(x)) \equiv 1 - c(x)\,. \tag{43'}$$

**Proof.** — Indeed (31) can be written:

$$\forall\, I_{j_1 \ldots j_{r-1}1}\,, \quad \sigma I_{j_1 \ldots j_{r-1}1} = I_{j'_1 \ldots j'_{r-1}1} \tag{31'}$$

with: $j_l + j'_l = 1$ for $l = 1, 2, \ldots, r-1$.

Being, therefore:

$$x \in I_{j_1 \ldots j_{r-1}1} \Longrightarrow \begin{cases} c(x) = \dfrac{j_1}{2} + \ldots + \dfrac{j_{r-1}}{2^{r-1}} + \dfrac{1}{2^r} \\[2ex] \sigma(x) \in I_{j'_1 \ldots j'_{r-1}1} \Longrightarrow c(\sigma(x)) = \dfrac{j'_1}{2} + \ldots + \dfrac{j'_{r-1}}{2^{r-1}} + \dfrac{1}{2^r}\,, \end{cases}$$

(43′) can be written [remember that $A = \bigcup I_{j_1 \ldots j_r}$]:

$$\forall x \in I_{j_1 \ldots j_{r-1}1}\,, \quad \frac{j'_1}{2} + \ldots + \frac{j'_{r-1}}{2^{r-1}} + \frac{1}{2^r} = 1 - \left(\frac{j_1}{2} + \ldots + \frac{j_{r-1}}{2^{r-1}} + \frac{1}{2^r}\right)\,. \tag{43''}$$

But (43″) is certainly true; in fact, being $j_l + j'_l = 1$ for $l = 1, 2, \ldots r-1$ (see (31′)), we get:

$$\left(\frac{j'_1}{2} + \ldots + \frac{j'_{r-1}}{2^{r-1}} + \frac{1}{2^r}\right) + \left(\frac{j_1}{2} + \ldots + \frac{j_{r-1}}{2^{r-1}} + \frac{1}{2^r}\right) =$$
$$= \left[\frac{1}{2} + \ldots + \frac{1}{2^{r-1}}\right] + \frac{2}{2^r} = \frac{1}{2}\left[\frac{1 - 1/2^{r-1}}{1/2}\right] + \frac{1}{2^{r-1}} = 1\,.$$

The assertion (43′) is proved. q.e.d.

Let us now consider two points $x, y \in A$ fulfilling conditions of the type:

$$x \in I_{j_1 \ldots j_{r-1}1}\,, \quad y \in I_{j_1 \ldots j_{r-1}0 j_{r+1} \ldots j_s-1 1}\,. \tag{44}$$

For the symmetrical points with respect to 1/2, thanks to (31) we get:

$$x' \equiv \sigma(x) \in I_{j'_1 \dots j'_{r-1} 1} \,, \quad y' \equiv \sigma(y) \in I_{j'_1 \dots j'_{r-1} 1 j'_{r+1} \dots j'_{s-1} 1} \,, \tag{44'}$$

whence the symmetrical points $x'$ and $y'$ fulfil (34). Consequently, by (42) and (43') we get:

$$\left. \begin{cases} x \in I_{j_1 \dots j_{r-1} 1} \,, \quad y \in I_{j_1 \dots j_{r-1} 0 j_{r+1} \dots j_{s-1} 1} \\ \\ \text{(i.e. } x \text{ and } y \text{ fulfil (44))} \end{cases} \right\} \Longrightarrow \tag{45}$$

$$\Longrightarrow |c(x) - c(y)| = |c(x') - c(y')| \le \left(\frac{2}{\delta^\lambda} + 1\right) |x' - y'|^\lambda \equiv \left(\frac{2}{\delta^\lambda} + 1\right) |x - y|^\lambda \,.$$

At this point we can enunciate the following :

**Lemma 2** — Given arbitrarily $\lambda \in ]0,1[$, and fixed consequently $k \in \mathbf{N}$ in such a way that $\lambda < (\log 2)/\log(2 + 1/k)$, the Cantor function (33) is $\lambda$-Hölder continuous in $A$; and therefore can be extended as a a $\lambda$-Hölder continuous function in [0,1].

**Proof.** Let $x$ and $y$ be two arbitrary points of $A$ and let, for example: $x < y$. Then, either $x$ and $y$ belong to the same interval $I_{j_1 \dots j_r}$ and hence $c(x) = c(y) = [0, j_1 \dots j_r]_b$ (see (33)) and trivially: $|c(x) - c(y)| \le |x - y|^\lambda$; or $x$ and $y$ belong to disjointed intervals (see (29''')) and the interval containing $x$ lies on the left of the interval containing $y$. But then one of the conditions (a) and (b) of Lemma 1 will be fulfilled. If condition (a) is fulfilled, we have:

$$x \in I_{j_1 \dots j_{s'}} \,, \quad y \in I_{j_1 \dots j_{s'} \dots j''_{s''}} \tag{a}$$

which is a condition of type (34); hence thanks to (42):

$$\text{if (a) is fulfilled, then: } |c(x) - c(y)| \le \left(\frac{2}{\delta^\lambda} + 1\right) |x - y|^\lambda \,. \tag{j}$$

If, on the contrary, condition (b) is fulfilled, we have:

(b) $$x \in I_{j_1 \dots j_{r-1} 0 j'_{r+1} \dots j'_{s'}} \,, \quad y \in \begin{cases} I_{j_1 \dots j_{r-1} 1} & (\text{if } s'' = r) \\ \\ I_{j_1 \dots j_{r-1} 1 j''_{r+1} \dots j''_{s''}} & (\text{if } s'' > r) \,. \end{cases}$$

In the first case ($s'' = r$) $x$ and $y$ fulfil (44); hence thanks to (45):

($j'$) if (b) is fulfilled with $s'' = r$ then : $|c(x) - c(y)| \leq \left(\frac{2}{\delta^\lambda} + 1\right) |x - y|^\lambda$ .

In the second case ($s'' > r$), fixed a point $z \in I_{j_1 \dots j_{r-1} 1}$, $y$ and $z$ fulfil (34), and thanks to (42) we have:

$$|c(y) - c(z)| \leq \left(\frac{2}{\delta^\lambda} + 1\right) |z - y|^\lambda \leq \left(\frac{2}{\delta^\lambda} + 1\right) |x - y|^\lambda \,,$$

while $x$ and $z$ fulfil (44), and thanks to (45) we have:

$$|c(x) - c(z)| \leq \left(\frac{2}{\delta^\lambda} + 1\right) |x - z|^\lambda \leq \left(\frac{2}{\delta^\lambda} + 1\right) |x - y|^\lambda \,.$$

Therefore:

($j''$) if (b) is fulfilled with $s'' > r$ then : $|c(x) - c(y)| \leq 2\left(\frac{2}{\delta^\lambda} + 1\right) |x - y|^\lambda$ .

In every case, whatever sub-condition is fulfilled, the implication is valid:

$$x < y \Longrightarrow |c(x) - c(y)| \leq 2\left(\frac{2}{\delta^\lambda} + 1\right) |x - y|^\lambda \,.$$

q.e.d.

## 5. Some Examples

From now on we consider $c(x)$ extended as the zero-function in the interval $]-\infty, 0]$. Obviously (see Lemma 2 and Proposition 4):

(46) $$\begin{cases} c(x) \text{ is } \lambda - \text{Hölder continuous in } ]-\infty, 1] \\ \\ c(x) \text{ is } \textit{not} \text{ absolutely continuous there.} \end{cases}$$

Furthermore we agree that the number $\lambda \in ]0,1[$ of Lemma 2 (and of (46)) be fixed $> 1 - \frac{1}{p}$ (where $p = 2 + 2(1+\alpha)(2-\beta)$ and $\alpha, \beta$ (and hence $p$) are fixed in correspondence with $\Lambda \equiv e$, [see (21)]; *do not confuse* $p \equiv 2+2(1+\alpha)(2-\beta)$ with the integer $p \equiv p(\lambda) \in \mathbf{N}$ of (38)); consequently we agree that the $k \in \mathbf{N}$ which intervenes in the construction of the open $A$ and of the Cantor function $c(x)$ be fixed in such a way that:

$$(47) \qquad 1-\frac{1}{p} \equiv 1-[2+2(1+\alpha)(2-\beta)]^{-1} < \lambda < (\log 2)/\log\left(2+\frac{1}{k}\right) .$$

The $\lambda$ of (46) is just that of (47).

At this point, fixed $x^i \equiv (x_1^i, \dots, x_5^i) \in ]1, 2\pi - 1[^5$ $(i = 1, \dots, 5)$, let us consider the function:

$$c_i(x_1, \dots, x_5) \equiv c_i(x) = c(\min\{1-|x_1-x_1^i|, \dots, 1-|x_5-x_5^i|\}), \quad x \in G \equiv ]0, 2\pi[^5,$$

$i = 1, \dots, 5$, ["Aztec Pyramid"].

Thanks to (46) and (47) we have:

$$(48) \quad \begin{cases} \forall\, x, y \in G\,, \ |c_i(x) - c_i(y)| \le 2\left(\dfrac{2}{\delta^\lambda}+1\right)|x-y|^\lambda \equiv K_0|x-y|^\lambda \quad \left(\lambda > 1-\dfrac{1}{p}\right) \\ \forall \text{ fixed } \hat{x}_k \in \displaystyle\prod_{j\neq k}^{1\le j\le 5} ]x_j^i - 1, x_j^i + 1[\,, \\ \hat{c}_i(x_k) \equiv c_i(\hat{x}_k, x_k) \ \text{ is \textit{not} absolutely continuous for } x_k \in [0, 2\pi]\,, \end{cases}$$

where:

$$\hat{x}_k = \begin{cases} (x_1, \dots, x_{k-1}, x_{k+1}, \dots, x_5) & , \quad \text{for } k = 2, 3, 4 \\ (x_2, \dots, x_5) & , \quad \text{for } k = 1 \\ (x_1, \dots, x_4) & , \quad \text{for } k = 5\,, \end{cases}$$

and moreover:

$$c_i(\hat{x}_k, x_k) = c_i(x_1, \dots, x_5) \ \text{ with } \begin{cases} \hat{x}_k \ \text{ fixed} \\ x_k \ \text{ variable in } [0, 2\pi]\,. \end{cases}$$

**Proposition 7** — The function $c_i(x)$ is $\lambda$-Hölder continuous in $G$ [and hence $BUC$ and hence $VMO$ in $G$]; moreover $c_i(x)$ belongs to $W_{\#}^{1-\frac{1}{p},p}(G)$ [where $p = 2+2(1+\alpha)(2-\beta)$]; but $c_i(x)$ *does not* belong to $W^{1,1}(G)$. [$G = ]0,2\pi[^5$].

**Proof.** — It is obvious that $c_i \notin W^{1,1}(G)$; since otherwise (see for example Nečas [Ne]) for almost all $\hat{x}_k \in ]0,2\pi[^4$ $(k = 1,\dots,5)$ the restriction $\hat{c}_i(x_k) \equiv c_i(\hat{x}_k, x_k)$ would be absolutely continuous in $[0,2\pi]$, which is prohibited by (48). That $c_i$ be $\lambda$-Hölder continuous in $G$ is assured by (48). And by (48) again it follows that $c_i \in W_{\#}^{1-\frac{1}{p},p}(G)$, since

$$\iint_{G\times G} \frac{|c_i(x) - c_i(y)|^p}{|x-y|^{5+(1-1/p)p}} dx\, dy \le K_0^p \int_G dx \int_G \frac{|x-y|^{\lambda p}}{|x-y|^{5+(1-\frac{1}{p})p}} dy < +\infty$$

being: $\lambda > 1 - \dfrac{1}{p}$. q.e.d.

Let us now consider, for all $i \in \{1,2,\dots,5\}$, an index $j \equiv j_i \ne i$ also belonging to $\{1,2,\dots,5\}$ and the function:

$$b_i(x_1,\dots,x_5) = \frac{(x_i - x_i^i)(x_j - x_j^i)}{(x_1 - x_1^i)^2 + \dots + (x_5 - x_5^i)^2}\,, \qquad x \in G\backslash\{x^i\}\,.$$

Obviously $b_i$ is the trace (for $t = 0$) of the function:

$$\mathcal{B}_i(x_1,\dots,x_5,t) = \frac{(x_i - x_i^i)(x_j - x_j^i)}{(x_1 - x_1^i)^2 + \dots + (x_5 - x_5^i)^2 + t^2}\,.$$

Since:

$$|\mathcal{B}_i| \le 1/2\,; \qquad \left|\frac{\partial \mathcal{B}_i}{\partial x_k}\right|\,, \quad \left|\frac{\partial \mathcal{B}_i}{\partial t}\right| \le \frac{4}{\{|x - x^i|^2 + t^2\}^{1/2}} \qquad (k = 1,\dots,5)$$

and $p = 2 + 2(1+\alpha)(2-\beta) < 6$, we have:

$$\mathcal{B}_i \in W^{1,p}(\Omega)\,; \quad i = 1,\dots,5\,; \quad \Omega = G\times]0,1[\,,$$

and consequently:

$$b_i \in W^{1-\frac{1}{p},p}(G)\,; \quad i = 1,\dots,5\,; \quad G = ]0,2\pi[^5\,. \tag{49}$$

Analogously, being:

$$|b_i| \le 1/2\,; \qquad \left|\frac{\partial b_i}{\partial x_k}\right| \le \frac{4}{|x - x^i|} \qquad (k = 1,\dots,5)$$

we have:

$$(49') \qquad b_i \in W^{1,4}(G) \; ; \quad i = 1, \ldots, 5 \; ; \quad G = ]0, 2\pi[^5 \; .$$

On the other hand, posed $B^i_\rho = \{x \in \mathbf{R}^5 : |x - x^i| \leq \rho\}$, for evident reasons of symmetry we get:

$$(b_i)_{B^i_\rho} \equiv \frac{1}{|B^i_\rho|} \int_{B^i_\rho} b_i(x) dx = 0 \; ,$$

and hence:

$$\int_{B^i_\rho} |b_i(x) - (b_i)_{B^i_\rho}| dx = \int_{B^i_\rho} \frac{|x_i - x^i_i| \cdot |x_j - x^i_j|}{|x - x^i|^2} dx \geq$$

$$\geq \int_{H^j_i \cap B^i_\rho} \frac{|x_i - x^i_i| \cdot |x_j - x^i_j|}{|x - x^i|^2} dx \; , \quad \text{where } H^j_i \equiv \left\{ x : x_j - x^i_j \geq \sqrt{\sum_{s \neq j} (x_s - x^i_s)^2} \right\}$$

Thus:

$$\frac{1}{|B^i_\rho|} \int_{B^i_\rho} |b_i(x) - (b_i)_{B^i_\rho}| dx \geq \frac{1}{|B^i_\rho|} \int_{H^j_i \cap B^i_\rho} \frac{|x_i - x^i_i|^2}{|x - x^i|^2} dx$$

$$= \frac{1}{|B^i_\rho|} \int_0^\rho r^4 dr \int_{H^j_0 \cap B_1^{\cdot}} \frac{r^2 \omega_i^2}{r^2} d\sigma(B_1^{\cdot}) = \frac{1}{|B_1| 5} \int_{H^j_0 \cap B_1^{\cdot}} \omega_i^2 d\sigma(B_1^{\cdot}) > 0 \; , \quad \forall \rho > 0 \; ,$$

where: $B_1 = \{x : |x|^2 \equiv x_1^2 + \ldots + x_5^2 \leq 1\}$ ,

$$B_1^{\cdot} = \{\omega : |\omega|^2 = \omega_1^2 + \ldots + \omega_5^2 = 1\} \; , \quad H^j_0 = \{x : x_j \geq \sqrt{\sum_{s \neq j} x_s^2}\} \; .$$

Consequently:

$$(49'') \qquad b_i \notin VMO \quad ; \qquad i = 1, \ldots, 5 \; .$$

**Proposition 8** — The functions $b_1, \ldots, b_5$ belong to $W^{1,4}(G)$ and also $W^{1-\frac{1}{p},p}(G)$; but *do not* belong to $VMO$.

If now $\zeta(z)$ is a function such that:

$$\zeta \in C_0^\infty(\mathbf{R}^5) \; ; \; \zeta(z) \equiv 0 \quad \text{for} \quad |z| \geq \frac{1}{4} \; ; \quad \zeta(z) \equiv 2e - 4 \quad \text{for} \quad |z| \leq \frac{1}{8} \; ,$$

from Proposition 7 and 8 it follows:

**Proposition 9** — With $p = 2 + 2(1 + \alpha)(2 - \beta)$ and $\lambda$ fulfilling (47), we have:

$$
(50) \qquad \begin{cases} c_i \notin W^{1,1}(G) \;, \quad c_i \in W_{\#}^{1-\frac{1}{p},p}(G) \;, \quad c_i \in C^{o,\lambda} \subset BUC \subset VMO \\ \\ \zeta(x-x^i)b_i \in W_{\#}^{1,4}(G) \cap W_{\#}^{1-\frac{1}{p},p}(G) \;, \quad \zeta(x-x^i)b_i \notin VMO \;. \end{cases}
$$

Well, let us at last consider the three following elliptic equations in $G \equiv ]0, 2\pi[^5$:

$$
(51) \qquad -\sum_i^{1...5}(1 + c_i(x))D^2_{x_i}w + w = g(x) \;,
$$

$$
(52) \qquad -\sum_i^{1...5}(2 + \zeta(x - x^i)b_i(x))D^2_{x_i}w + w = g(x) \;,
$$

$$
(53) \qquad -\sum_i^{1...5}(1 + c_i(x) + \zeta(x - x_i)b_i(x))D^2_{x_i}w + w = g(x) \;;
$$

On the basis of Proposition 9 we obtain:

**Proposition 10** — Equation (51) cannot be dealt neither with Theorem $\mathcal{M}_1$ nor with Theorem $\mathcal{M}_2$ of C. Miranda; but it can be dealt with Schauder Theory, as well with $VMO$ Theory, as well with the existence Theorem of [Cn2]. Equation (52) cannot be dealt with $VMO$ Theory; but it can be dealt with Theorem $\mathcal{M}_2$ of C. Miranda as well with the Existence Theorem [Cn2]. Equation (53) cannot be dealt neither with $VMO$ Theory, nor with Theorem $\mathcal{M}_2$ of C. Miranda; but it can be dealt with the existence Theorem of [Cn2].

**Remark.** — If the basis of the Aztec Pyramids $c_i$ are two by two disjointed, then equation (53) cannot be dealt even with Cordes Theory.

Indeed, called $a_i$ the coefficients of (53) [and posed $b_i(x^i) = 1/2$], in $x^1$ we have: $a_1(x^1) = e$, $a_i(x^1) = 1$ for $i \neq 1$; hence in $x^1$ (12) cannot be fulfilled, because in $x^1$:

$$
\sum_1^5 a_i^2 = 4 + e^2 > \frac{1}{4}(4+e)^2 = \frac{1}{5-1}\left(\sum_1^5 a_i\right)^2 \quad !
$$

## References


[Cn1 ] A. Canfora, Extension of an integral inequality of C. Miranda and applications to the elliptic equations with discontinuous coefficients, *Czechoslovak Math. Journal* **39** (1989) pp.385-421.

[Cn2 ] A. Canfora, Su una classe di equazioni ellittiche a coefficienti non VMO, *Rendiconti dell'Accademia delle Scienze Fisiche e Matematiche* (della Società Nazionale di Scienze Lettere e Arti, in Napoli), ser.IV, **LIX** CXXXI year (1992) pp.121-128.

[CnZ ] A. Canfora, P. Zecca, Su di un'equazione ellittica in $R^3$ a coefficienti discontinui e non derivabili, *Ann. Mat. Pura e Appl.* (4) **141** (1985) pp.369-384.

[CFL1 ] F. Chiarenza, M. Frasca, P. Longo, Interior $W^{2,p}$ estimates for non divergence elliptic equations with $VMO$ coefficients, *Ric. di Mat.*, **XL**, nr.1 (1991), pp.149-168.

[CFL2 ] F. Chiarenza, M. Frasca, P. Longo, $W^{2,p}$ solvability of the Dirichlet problem for non divergence elliptic equations with $VMO$ coefficients.

[DG ] E. De Giorgi, Sulla differenziabilità e l'analiticità delle estremali degli integrali multipli regolari, *Mem. Accad. Sci.*, *Torino*, Cl. Sci. Fis. Mat. Natur. (3) **3** (1957) pp.25-43.

[Gg ] E. Gagliardo, Ulteriori proprietà di alcune classi di funzioni in più variabili, *Ric. di Mat.*, Napoli **8** (1959) pp.24-51.

[GT ] D. Gilbarg, N.S. Trudinger, Elliptic partial differential equations of second order, $2^{nd}$ edition, Springer-Verlag, Berlin, 1983.

[KS1 ] N.V. Krylov, M.V. Safonov, An estimate of the probability that a diffusion process hits a set of positive measure, *Dokl. Akad. Nauk SSSR* **245** (1979) pp.253-255 (Russian). English transl. in *Soviet Math. Dokl.* **20** (1979) pp.253-255.

[KS2 ] N.V. Krylov, M.V. Safonov, Certain properties of solutions of parabolic equations with measurable coefficients, *Izvestija Akad. Nauk SSSR* **40** (1980) pp.161-175 (Russian). English transl. in *Math. USSR Izvestija* **16** (1981) n.1 pp.151-164.

[M1 ] C. Miranda, Sulle equazioni ellittiche del secondo ordine a coefficienti discontinui, *Ann. Mat. Pura e Appl.* **63** (1963) pp.353-386.

[M2 ] C. Miranda, Su alcune diseguaglianze integrali, *Memorie Accad. Lincei* (Sc. Fisiche), ser.VIII, **VII**, sez.1, 1 (1963) pp.1-14.

[N1 ] L. Nirenberg, On elliptic partial differential equations (Lecture II), *Ann. Sc. Norm. Sup. Pisa* **13** (1959) pp. 123-131.

[N2 ] L.Nirenberg, An extended interpolation inequality, *Ann. Sc. Norm. Sup. Pisa* **20** (1966) pp.733-737.

[Ne ] J. Nečas, Les Méthodes directes en Théorie des Équations Elliptiques, Masson et Cie Éditeurs, Paris.

[Z1 ] P. Zecca, Teoremi di esistenza e di unicità di soluzioni periodiche per una equazione ellittica a coefficienti discontinui e non derivabili, *Ric. di Mat.* **35** (1986) pp.61-74.

[Z2 ] P. Zecca, Un nuovo teorema di esistenza per una nuova equazione ellittica a coefficienti discontinui e non derivabili, *Ric. di Mat.* **35** (1986) pp.215-219.

Albino Canfora

Dipartimento di Matematica and Applicazioni "R. Caccioppoli"

Università degli Studi di Napoli "Federico II"

Compl. universitario Monte S.Angelo - Via Cintia

80126 Napoli - Italy

P. CANNARSA, G. DA PRATO

# Schauder estimates for Kolmogorov equations in Hilbert spaces

## 1. Introduction

We are here concerned with an elliptic equation in a separable Hilbert space $H$ :

$$\mu\varphi(x) - \frac{1}{2}\operatorname{Tr}[D_x^2\varphi(x)] - < Bx, D_x\varphi(x) > = g(x), \; x \in H, \quad \mu > 0, \tag{1.1}$$

under the following assumptions.

**Hypothesis 1.**

*(i)* *$B : D(B) \subset H \to H$ is the infinitesimal generator of a strongly continuous semigroup $e^{tB}$ in $H$.*

*(ii)* *For any $t > 0$ the operator $Q_t$, defined by*

$$Q_t x = \int_0^t e^{tB} e^{tB^*} x dt, \quad x \in H,$$

*is trace class.*

*(iii)* *The mapping $g : H \to \mathrm{R}$, is uniformly continuous and bounded.*

Equation (1.1) is connected with the differential stochastic Ito equation

$$\begin{cases} dX(t) = BX(t)dt + dW_t \\ \\ X(0) = x, \end{cases} \tag{1.2}$$

where $W(t)$ is a cylindrical process in $H$. It is well known, see e.g. [6], that, under Hypothesis 1, equation (1.2) has a unique mild solution $X(t, x)$, $t \geq 0$, given by

$$X(t, x) = e^{tB}x + W_B(t),$$

where

$$W_B(t) = \int_0^t e^{(t-s)B} dW(s).$$

Let us introduce the *Markov transition semigroup* on $C_b(H)$.

$$P_t\varphi(x) = \mathbb{E}\left[\varphi(X(t,x))\right],\ t \geq 0,\ x \in H,\ \varphi \in C_b(H). \tag{1.3}$$

A natural candidate for the solution of (1.1) is given by

$$\varphi(x) = (F(\mu)g)(x) = \int_0^\infty e^{-\mu t} P_t g(x)dt,\ x \in H. \tag{1.4}$$

We recall that the semigroup $P_t,\ t \geq 0$ is weakly continuous, in particular that the mapping

$$[0,+\infty[\to \mathbb{R},\ t \to P_t\varphi(x)$$

is continuous for all $\varphi \in C_b(H)$ and all $x \in H$, see [3]. In the same paper it is proved that $F(\cdot)$ is a pseudo–resolvent in $C_b(H)$ and it is one–to–one. Consequently, there exists a linear operator

$$\mathcal{L} : D(\mathcal{L}) \subset C_b(H) \to C_b(H),$$

such that

$$F(\mu) = (\mu - \mathcal{L})^{-1},\ \mu > 0.$$

Thus the function $\varphi$, given by (1.4), belongs to $D(\mathcal{L})$ and it can be considered as a generalized solution of equation (1.1) in the following sense, see [4]: there exists a sequence $\{\varphi_n\} \subset D(\mathcal{L})$, uniformly bounded in $H$, such that $\varphi_n(x) \to \varphi(x)$ and $\mu\varphi_n(x) - \mathcal{L}\varphi_n(x) \to g(x)$, for any $x \in H$.

In this paper we want to prove a regularity property of $\varphi$, namely that if $g$ is Hölder continuous then $\varphi$ is twice differentiable and $D_x^2\varphi$ is also Hölder continuous with the same exponent of $g$. When $H$ is finite–dimensional this result was proved in [5]. The main ingredients for the proof are a recent interpolation result, [2], see Theorem 2 below, an interpolatory argument due to [7], and some estimates on the derivatives of the transition semigroup that we recall in the following proposition.

**Proposition 1.** *Assume that Hypothesis* 1 *holds, and let $\varphi \in C_b(H)$. Then, $P_t\varphi$ is infinitely many times differentiable and for all $n \in \mathbb{N}$ there exists $C_n > 0$ such that*

$$\|D_x^n P_t\varphi\|_0 \leq C_n t^{-\frac{n}{2}} \|\varphi\|_0,\ \forall\, \varphi \in C_b(H),\ n \in \mathbb{N} \cup \{0\}, \tag{1.5}$$

$$\|D_x^n P_t\varphi\|_0 \leq C_n t^{-\frac{n-1}{2}} \|D\varphi\|_0,\ \forall\, \varphi \in C_b^1(H),\ n \in \mathbb{N}. \tag{1.6}$$

Proposition 1 can be proved both by probabilistic, see e.g. [6], and by analytic tools, see [2].

In §2 we recall some results about interpolation spaces that will be needed and in §3 we prove Schauder estimates.

Let us give some notation. We denote by $C_b(H)$ the Banach space of all uniformly continuous and bounded functions $\varphi : H \to \mathbb{R}$, endowed with the norm $\|\varphi\|_0 = \sup_{x\in H} |\varphi(x)|$. For any $k \in \mathbb{N}$ we denote by $C_b^k(H)$ the subset of $C_b(H)$ of all functions that are uniformly continuous and bounded together with all their derivatives of order less or equal to $k$. For any $\varphi \in C_b^k(H)$ we set

$$[\varphi]_k = \sup_{x\in H} \|D^k\varphi(x)\|,$$

and

$$\|\varphi\|_k = \|\varphi\|_0 + [\varphi]_k.$$

For any $\theta \in ]0,1[$ we shall denote by $C_b^\theta(H)$ the subspace of $C_b(H)$ consisting of all functions $\varphi$ that are $\theta$–Hölder continuous, that is such that

$$|\varphi|_\theta := \sup_{x,y\in H, x\neq y} \frac{|\varphi(x)-\varphi(y)|}{|x-y|^\theta} < +\infty.$$

Moreover, for all $k \in \mathbb{N}$, $C_b^{k+\theta}(H)$ will represent the set of all functions of $C_b^k(H)$ that have their $k^{th}$ derivative $\theta$–Hölder continuous.

## 2. Interpolation spaces

Let $X, Y$ be Banach spaces such that $Y \subset X$ with continuous embedding. We set

$$K(t,x) = \inf\{\|a\|_X + t\|b\|_Y : x = a+b, a \in X, b \in Y\},$$

and, for any $\theta \in ]0,1[$

$$(X,Y)_{\theta,\infty} = \{x \in X : \|x\|_{(X,Y)_{\theta,\infty}} := \sup_{t>0} \frac{1}{t^\theta} K(t,x) < +\infty\}.$$

$(X,Y)_{\theta,\infty}$ is a Banach space with the norm $\|\cdot\|_{(X,Y)_{\theta,\infty}}$.

From the definition above it follows that $x \in (X,Y)_{\theta,\infty}$ provided

$$x = a(t) + b(t),\ \ a(t) \in X,\ b(t) \in Y,\ t > 0,$$

and, for some constant $C > 0$,

$$\|a(t)\|_X \le Ct^\theta,\ \|b(t)\|_X \le Ct^{\theta-1},\ t>0.$$

We recall the basic interpolation theorem, see e.g. [9].

**Theorem 1.** *Let $X, X_1, Y, Y_1$ be Banach spaces such that*

$$Y \subset X,\ Y_1 \subset X_1$$

*with continuous embeddings. Let moreover $T$ be a linear application*

$$T : X \to X_1,\ \ T : Y \to Y_1,$$

*such that*

$$\|Tx\|_{X_1} \le M_X \|x\|_X,\ \forall\, x \in X,\ \ \|Ty\|_{Y_1} \le M_Y \|y\|_Y, \forall\, y \in Y.$$

*Then we have*

$$\|Tx\|_{(X_1,Y_1)_{\theta,\infty}} \le M_X^{1-\theta} M_Y^{\theta}\ \|x\|_{(X,Y)_{\theta,\infty}},\ \forall\, x \in (X,Y)_{\theta,\infty}.$$

The following interpolation result was proved in [2].

**Theorem 2.** *For any $\theta \in ]0,1[$ we have*

$$(C_b(H), C_b^1(H))_{\theta,\infty} = C_b^{\theta}(H).$$

We will need also the following inclusion result.

**Proposition 3.** *Let $\alpha, \theta \in ]0,1[$. Then we have*

$$(C_b^2(H), C_b^{2+\alpha}(H))_{\theta,\infty} \subset C_b^{2+\alpha\theta}(H).$$

**Proof** — Let $\varphi \in (C_b^2(H), C_b^{2+\alpha}(H))_{\theta,\infty}$. Then there exist $a(t,\cdot),\ b(t,\cdot)$ such that

$$\varphi = a(t,\cdot) + b(t,\cdot),\ t > 0,$$

with

$$\|D^2 a(t,\cdot)\|_0 \le Ct^{\theta},\ \ \|D^2 b(t,\cdot)\|_{\alpha} \le Ct^{\theta-1},\ t > 0.$$

We have

$$\begin{aligned} \|D^2\varphi(x) - D^2\varphi(y)\| &\le \|D^2 a(t,x) - D^2 a(t,y)\| + \|D^2 b(t,x) - D^2 b(t,y)\| \\ &\le 2Ct^{\theta} + C|x-y|^{\alpha} t^{\theta-1}. \end{aligned}$$

Setting $t = |x-y|^{\alpha}$ the conclusion follows. ■

### 2.1. The class $J_\alpha(X,Y)$

Let $\alpha \in [0,1]$ and let $E$ be another Banach space such that $Y \subset E \subset X$ with continuous embeddings.

We say that $E$ belongs to the class $J_\alpha(X,Y)$ if there exists $C > 0$ such that

$$\|x\|_E \le C\|x\|_X^{1-\alpha}\, \|x\|_Y^{\alpha},\ \forall\, x \in Y.$$

Obviously

$$X \in J_0(X,Y),\quad Y \in J_1(X,Y).$$

Moreover it is easy to see that

$$(X,Y)_{\theta,\infty} \in J_\theta(X,Y),\ \forall\, \theta \in ]0,1[.$$

Now we recall the Reiteration Theorem. For a proof see for instance [8]

**Theorem 4.** *Let $Y, F, E, X$ be Banach spaces such that*

$$Y \subset F \subset E \subset X,$$

*with continuous embeddings.*

*Assume that $E \in J_\alpha(X,Y)$ and $F \in J_\beta(X,Y)$ for some $\alpha, \beta \in [0,1]$, with $\alpha \neq \beta$. Then, for all $\theta \in ]0,1[$*

$$(E,F)_{\theta,\infty} \supset (X,Y)_{(1-\theta)\alpha+\theta\beta,\infty}, \tag{2.1}$$

*with continuous inclusion.*

*Moreover, if $E \subset (X,Y)_{\alpha,\infty}$ and $F \subset (X,Y)_{\beta,\infty}$, for some $\alpha, \beta \in ]0,1[$ with $\alpha \neq \beta$. Then for all $\theta \in ]0,1[$,*

$$(X,Y)_{(1-\theta)\alpha+\theta\beta,\infty} \supset (E,F)_{\theta,\infty}. \tag{2.2}$$

Now we can characterize interpolations spaces between $C_b(H)$ and $C_b^\alpha(H)$ and between $C_b^\alpha(H)$ and $C_b^1(H)$.

**Proposition 5.** *Let $\beta, \theta \in ]0,1[$. Then we have*

$$\left(C_b(H), C_b^\beta(H)\right)_{\theta,\infty} = C_b^{\theta\beta}(H), \tag{2.3}$$

and

$$\left(C_b^\beta(H), C_b^1(H)\right)_{\theta,\infty} = C_b^{\beta+\theta(1-\beta)}(H). \tag{2.4}$$

**Proof** — To prove (2.3) it is enough to apply Theorem 4, with

$$Y = C_b^1(H),\ F = C_b^\beta(H),\ E = X = C_b(H),\ \alpha = 0,$$

and recall Theorem 2. The proof of (2.4) is similar. ∎

## 2.2. Interpolatory estimates

This subsection is devoted to proving interpolatory inclusions that we will need in the sequel. They are well known when $H$ is finite dimensional.

**Proposition 6.** *We have*

$$C_b^1(H) \in J_{1/2}\left(C_b(H), C_b^2(H)\right) \tag{2.5}$$

**Proof** — Let $u \in C_b^2(H)$ and let $h > 0$, $x, z \in H$ with $|z| = 1$. Then, from the equality

$$u(x+hz) = u(x) + hDu(x) \cdot z + h^2 \int_0^1 (1-\sigma) < D^2u(x+\sigma hz) \cdot z, z > d\sigma,$$

it follows

$$|Du(x) \cdot z| \leq \frac{2}{h} \|u\|_0 + \frac{h}{2} \|D^2u\|_0, \ \forall \, h > 0.$$

Since

$$\min_{h>0} \left(\frac{a}{h} + bh\right) = 2\sqrt{ab},$$

we have

$$|Du(x) \cdot z| \leq 2\|u\|_0^{1/2} \, \|D^2u\|_0^{1/2}.$$

■

**Proposition 7.** *For any $\alpha \in ]0,1[$ we have*

$$C_b^2(H) \in J_{\frac{1}{1+\alpha}}\left(C_b^1(H), C_b^{2+\alpha}(H)\right), \tag{2.6}$$

*and*

$$\|u\|_2 \leq C\|D^2u\|_\alpha^{\frac{1}{1+\alpha}} \ \|u\|_1^{\frac{\alpha}{1+\alpha}} \tag{2.7}$$

**Proof** — Let $u \in C_b^{2+\alpha}(H)$ and let $h > 0$, $x, z \in H$ with $|z| = 1$. Then, from the equality

$$u(x+hz) = u(x) + hDu(x) \cdot z + \frac{1}{2}h^2 < D^2u(x) \cdot z, z >$$

$$+h^2 \int_0^1 (1-\sigma) < [D^2u(x+\sigma hz) - D^2u(x)] \cdot z, z > d\sigma,$$

it follows

$$\frac{1}{2}h^2| < D^2u(x)\cdot z, z > | \le |u(x+hz) - u(x) - hDu(x)\cdot z|$$

$$+\frac{1}{2}h^2\|D^2u\|_\alpha \int_0^1 (1-\sigma)\sigma^\alpha h^\alpha \, d\sigma$$

$$\le 2h\|Du\|_0 + C_\alpha h^{2+\alpha}\|D^2u\|_\alpha.$$

So

$$\|D^2u(x)\| \le \frac{2}{h}\|u\|_1 + C_\alpha h^\alpha \|D^2u\|_\alpha.$$

The conclusion follows taking minimum for $h > 0$. ∎

**Proposition 8.** *For any $\alpha \in ]0,1[$ we have*

$$C_b^1(H) \in J_{\frac{1-\alpha}{2-\alpha}}\left(C_b^\alpha(H), C_b^2(H)\right), \tag{2.8}$$

*and*

$$\|u\|_1 \le C_\alpha \|u\|_\alpha^{\frac{1}{2-\alpha}} \; \|u\|_2^{\frac{1-\alpha}{2-\alpha}} \tag{2.9}$$

**Proof** — Let $u \in C_b^{2+\alpha}(H)$ and let $h > 0$, $x, z \in H$ with $|z| = 1$. Then, from the equality

$$u(x+hz) = u(x) + hDu(x)\cdot z + h^2 \int_0^1 (1-\sigma) < D^2u(x+\sigma hz)\cdot z, z > d\sigma,$$

it follows

$$\|Du\|_0 \le h^{\alpha-1}\|u\|_\alpha + h^2\|D^2u\|_0.$$

The conclusion follows taking minimum for $h > 0$. ∎

**Proposition 9.** *For any $\alpha \in ]0,1[$ we have*

$$C_b^2(H) \in J_{1-\alpha/2}\left(C_b^\alpha(H), C_b^{2+\alpha}(H)\right), \tag{2.10}$$

*and*

$$\|u\|_2 \le C_\alpha \|u\|_{2+\alpha}^{1-\alpha/2} \; \|u\|_\alpha^{\alpha/2} \tag{2.11}$$

**Proof** — By (2.7) we have

$$\|u\|_2 \le C_\alpha \|u\|_{2+\alpha}^{\frac{1}{1+\alpha}} \ \|u\|_1^{\frac{\alpha}{1+\alpha}},$$

then taking into account (2.9) it follows

$$\|u\|_2 \le C_1 \|u\|_{2+\alpha}^{\frac{1}{1+\alpha}} \ \|u\|_\alpha^{\frac{\alpha}{(2-\alpha)(1+\alpha)}} \ \|u\|_2^{\frac{\alpha(1-\alpha)}{(2-\alpha)(1+\alpha}\big)},$$

which yields

$$\|u\|_2^{\frac{2}{(1+\alpha)(2-\alpha)}} \le C_2 \|u\|_{2+\alpha}^{\frac{1}{1+\alpha}} \ \|u\|_\alpha^{\frac{\alpha}{(2-\alpha)(1+\alpha)}},$$

that implies (2.11). ■

The main result of this subsection is the following inclusion.

**Theorem 10.** *Let $\alpha, \gamma \in ]0,1[$. Then we have*

$$C_b^{2+\alpha\gamma} \supset \left(C_b^\alpha(H), C_b^{2+\alpha}(H)\right)_{1-\frac{\alpha}{2}(1-\gamma),\infty}, \tag{2.12}$$

*with continuous embedding.*

**Proof** —

We first prove the inclusion

$$\left(C_b^2(H), C_b^{2+\alpha}(H)\right)_{\gamma,\infty} \supset \left(C_b^\alpha(H), C_b^{2+\alpha}(H)\right)_{1-\frac{\alpha}{2}(1-\gamma),\infty}. \tag{2.13}$$

We apply the Reiteration Theorem with

$$Y = C_b^{2+\alpha}(H), \ F = C_b^{2+\alpha}(H), \ E = C_b^2(H), \ X = C_b^\alpha(H).$$

Then

$$E \in J_{1-\alpha/2}(X,Y), \quad F \in J_1(X,Y).$$

So, by Theorem 4

$$(E,F)_{\gamma,\infty} \supset (X,Y)_{(1-\gamma)(1-\alpha/2),\infty},$$

and (2.13) is proved. Now the conclusion follows from Proposition 3. ■

## 3. Schauder estimates

We now prove the main result of the paper.

**Theorem 1.** *Assume that Hypothesis* 1 *holds. Let* $g \in C_b^\theta(H)$, $\theta \in ]0,1/2[$, *and let* $\varphi$ *be defined by* (1.4). *Then* $\varphi \in C_b^{2+\theta}(H)$.

**Proof** — We will use the interpolation argument of [7].

**Step 1**— For any $\theta \in ]0,1[$ and for any $\alpha \in ]\theta,1[$, there exists $C_{\alpha,\theta} > 0$ such that

$$\|P_t g\|_\alpha \le C_{\alpha,\theta} t^{-\frac{\alpha-\theta}{2}} \|g\|_\theta, \ \forall\, g \in C_b^\theta(H). \tag{3.1}$$

We first remark that, in view of (1.5) and (1.6), we have for $t > 0$,

$$P_t \in \mathcal{L}(C_b(H), C_b(H)) \cap \mathcal{L}(C_b(H), C_b^1(H)).$$

Thus, by Theorems 1 and 2 it follows that for any $\beta \in ]0,1[$,

$$P_t \in \mathcal{L}(C_b(H), C_b^\beta(H))$$

and there exists $C_\beta > 0$ such that

$$\|P_t g\|_\beta \le t^{-\frac{\beta}{2}} \|g\|_0, \ \forall\, g \in C_b(H).$$

On the other hand we have clearly

$$\|P_t g\|_1 \le \|g\|_1, \ \forall\, g \in C_b^1(H).$$

So, again by interpolation, recalling Proposition 5, we find that for any $\theta \in ]0,1[$ there exists $C_{\beta,\theta} > 0$ such that

$$\|P_t g\|_{\beta+\theta(1-\beta)} \le C_{\beta,\theta} t^{-\frac{\beta(1-\theta)}{2}} \|g\|_\theta, \ \forall\, g \in C_b^\theta(H).$$

Now it is enough to choose $\beta$ such that $\alpha = \beta + \theta(1-\beta)$.

**Step 2**— For any $t > 0$ there exists $C^1_{\alpha,\theta} > 0$ such that

$$\|D_x^2 P_t g\|_\alpha \le C^1_{\alpha,\theta} t^{\frac{\theta}{2}-\frac{\alpha}{2}-1} \|g\|_\theta, \ t > 0, \ \forall\, g \in C_b^\theta(H). \tag{3.2}$$

By estimates (1.5) and (1.6) we have

$$\begin{cases} \|P_t g\|_2 \le \dfrac{C_2}{t} \|g\|_0, \ \forall\, g \in C_b(H), \\[2ex] \|P_t g\|_2 \le \dfrac{C_2}{\sqrt{t}} \|g\|_1, \ \forall\, g \in C_b^1(H). \end{cases}$$

By using Theorem 1 and Theorem 2 we find

$$\|D_x^2 P_t g\|_0 \le C_2 t^{\frac{\theta}{2}-1}\|g\|_\theta,\ t>0,\ \forall\, g\in C_b^\theta(H). \tag{3.3}$$

Moreover, again by estimates (1.5) and (1.6) we have

$$\begin{cases} \|P_t g\|_3 \le \dfrac{C_3}{t^{3/2}}\|g\|_0,\ \forall\, g\in C_b(H) \\ \\ \|P_t g\|_3 \le \dfrac{C_3}{t}\|g\|_1,\ \forall\, g\in C_b^1(H). \end{cases}$$

By Theorem 1 it follows

$$\|D_x^3 P_t g\|_0 \le C_3 t^{\theta/2-3/2}\|g\|_\theta,\ t>0,\ \forall\, g\in C_b^\theta(H). \tag{3.4}$$

Fix now $t>0$ and define for any $h\in H$

$$T_h g =< D^2 P_t g\cdot h, h>$$

By (3.3) we have $T_h \in \mathcal{L}(C_b^\theta(H), C_b(H))$ and

$$\|T_h g\|_0 \le C_2 t^{\frac{\theta}{2}-1}\|g\|_\theta |h|^2.$$

Moreover by (3.4) we have $T_h\in\mathcal{L}(C_b^\theta(H), C_b^1(H))$ and

$$\|DT_h g\|_0 \le C_3 t^{\theta/2-3/2}\|g\|_\theta |h|^2.$$

By Theorem 2 there exists $C_{\alpha,\theta}^1>0$ such that

$$\|T_h g\|_\alpha \le C_{\alpha,\theta}^1 t^{\frac{\theta}{2}-\frac{\alpha}{2}-1}\|g\|_\theta |h|^2.$$

The conclusion follows now by the arbitraryness of $h$.

**Step 3**— Conclusion.

Let $\varphi$ be defined by (1.4). We first remark that, by using (3.3), one can show that $\varphi\in C_b^2(H)$.

We now prove that

$$\varphi\in \left(C_b^\alpha(H), C_b^{2+\alpha}(H)\right)_{1-\frac{\alpha}{2}+\frac{\theta}{2},\infty}. \tag{3.5}$$

Since by Theorem 10 with $\alpha\gamma=\theta$ we have

$$\left(C_b^\alpha(H), C_b^{2+\alpha}(H)\right)_{1-\frac{\alpha}{2}+\frac{\theta}{2},\infty}\subset C_b^{2+\theta}(H), \tag{3.6}$$

the conclusion will follow.

To prove (3.5) we set

$$\varphi(x) = a(t,x) + b(t,x),$$

where

$$a(t,x) = \int_0^t e^{-\mu s} P_s g(x) ds,$$

and

$$b(t,x) = \int_t^{+\infty} e^{-\mu s} P_s g(x) ds.$$

Then from (3.1) it follows that

$$\begin{aligned} \|a(\cdot,t)\|_\alpha &\leq C_{\alpha,\theta} \int_0^t e^{-\mu s} s^{-\frac{\alpha-\theta}{2}} \, ds \; \|g\|_\theta \\ &= C_{\alpha,\theta} t^{1-\frac{\alpha-\theta}{2}} \int_0^1 e^{-\mu t\sigma} \sigma^{-\frac{\alpha-\theta}{2}} \, d\sigma \; \|g\|_\theta \\ &\leq \frac{C_{\alpha,\theta}}{1-\frac{\alpha-\theta}{2}} t^{1-\frac{\alpha-\theta}{2}} \|g\|_\theta, \end{aligned}$$

and from (3.2) that

$$\begin{aligned} \|b(\cdot,t)\|_{2+\alpha} &\leq C^1_{\alpha,\theta} \int_t^{+\infty} e^{-\mu s} s^{-\frac{\alpha-\theta}{2}-1} \, ds \; \|g\|_\theta \\ &= C^1_{\alpha,\theta} t^{-\frac{\alpha-\theta}{2}} \int_1^{+\infty} e^{-\mu t\sigma} \sigma^{-\frac{\alpha-\theta}{2}-1} \, d\sigma \; \|g\|_\theta \\ &\leq \frac{C^1_{\alpha,\theta}}{\alpha-\theta} t^{\frac{\theta-\alpha}{2}} \|g\|_\theta, \end{aligned}$$

This implies (3.5). ∎

## References


[1] CANNARSA P. & DA PRATO G. (1993) *On a Functional Analysis approach to Parabolic Equations in infinite dimensions,* J. Functional Analysis, vol. 118, n. 1, 22–42.

[2] CANNARSA P. & DA PRATO G. (1994) *An interpolation result for Hilbert space valued continuous functions*, preprint S.N.S. n.27. .

[3] CERRAI S. (1994) *A Hille-Yosida theorem for weakly continuous semigroups*, Semigroup Forum, Vol. 49, 349–367.

[4] CERRAI S. & GOZZI F. (1994) *Strong solutions of Cauchy problems associated to weakly continuous semigroups,* Preprint S.N.S. n.35.

[5] DA PRATO G. & LUNARDI A. (to appear) *On the Ornstein-Uhlenbeck operator in spaces of continuous functions,* J. Functional Analysis.

[6] DA PRATO G. & ZABCZYK J. (1992) STOCHASTIC EQUATIONS IN INFINITE DIMENSIONS. Encyclopedia of Mathematics and its Applications, Cambridge University Press.

[7] LUNARDI A. (1994) *Schauder estimates for a class of degenerate elliptic and parabolic operators with unbounded coefficients in* $\mathbb{R}^n$., preprint S.N.S. n.30.

[8] LUNARDI A. (1995) ANALYTIC SEMIGROUPS AND OPTIMAL REGULARITY IN PARABOLIC PROBLEMS, Birkäuser, Basel.

[9] TRIEBEL H. (1986) *Interpolation Theory, Function Spaces, Differential Operators,* North-Holland, Amsterdam.

Piermarco Cannarsa

Dipartimento di Matematica, Università di Roma "Tor Vergata"

Via della Ricerca Scientifica, 00133, Roma, Italy

Giuseppe Da Prato

Scuola Normale Superiore di Pisa

Piazza dei Cavalieri 6, 56126 Pisa, Italy

M.C. CERUTTI, E.B. FABES, P. MANSELLI

# Uniqueness for elliptic equations with time-independent coefficients

## 1. Introduction

Throughout the paper let $z = (x, t) = (x_1, x_2, ..., x_d, t) \in \mathbf{R}^{d+1}$ and let $\Omega \subset \mathbf{R}^{d+1}$ $(d \geq 2)$ be a smooth domain with boundary $\partial\Omega$. (We'll refer to $x$ as the space variable and to $t$ as the time variable.) Given the elliptic operator

$$L = \sum_{h,j=1}^{d+1} a^{hj}(z) \frac{\partial^2}{\partial z_h \partial z_j} \tag{1.1}$$

where the coefficients $a^{ij}$ satisfy the uniform ellipticity condition

$$0 < \lambda|\xi|^2 \leq \sum_{h,j=1}^{d} a^{hj}(x)\xi_h\xi_j \leq \Lambda|\xi|^2 \tag{1.2}$$

the full Dirichlet problem for $L$ in $\Omega$ consists in finding a function $u(z)$, for $z \in \Omega$ satisfying

$$\begin{cases} Lu = f & \text{in } \Omega \\ u|_{\partial\Omega} = g \, , \end{cases} \tag{1.3}$$

with $f$ and $g$ given and even smooth functions defined on $\Omega$ and $\partial\Omega$, respectively.

A well developed theory of existence, uniqueness, and regularity of solutions for the full Dirichlet problem exists when the coefficient matrix $\left(a_{ij}(z)\right)$ is smooth in $\overline{\Omega}$ or even only continuous in $\overline{\Omega}$. However, when we assume only bounded and measurable coefficients, then a well understood theory is not available.

One notion of solution to (1.3) in the case of nonsmooth coefficients is that of good solution, that has been given in [4] and, with slight modifications, in [18] and [10].

**Definition 1.1** A function $u \in C^0(\overline{\Omega})$ is a *good solution* to (1.3) if

(i) there exist a sequence of operators

$$L_n = \sum_{h,j=1}^{d+1} a_n^{hj}(z) \frac{\partial^2}{\partial z_h \partial z_j}$$

defined on $\Omega$ with coefficients $a_n^{hj}$ satisfying the same ellipticity condition (1.2) as $L$;

(ii) $a_n^{hj}$ are smooth in $\overline{\Omega}$ and $a_n^{hj} \to a^{hj}$ a. e. in $\Omega$ as $n \to \infty$;

(iii) there exists a sequence of functions $u_n$ smooth in $\Omega$ that solve (uniquely)

$$\begin{cases} L_n u_n = f & \text{in } \Omega \\ u_n|_{\partial\Omega} = g \,, \end{cases} \tag{1.4}$$

and such that $u_n \to u$ uniformly in $\overline{\Omega}$.

The results of Krylov and Safonov on the Harnack principle and apriori Hölder continuity of the solution $u_n$ of (1.4) ([11], [16]) imply that the sequence $\{u_n\}$ is an equicontinuous family and hence a subsequence converges uniformly in $\overline{\Omega}$. Therefore a *good* solution to (1.3) always exists.

Concerning regularity of *good* solutions, Safonov showed by an example that one cannot expect more than a global Hölder continuity depending on the upper and lower bounds of the eigenvalues of the matrix $\big(a_{ij}(z)\big)$ ([17]). At the same time a *good* solution has a second order differential at almost every point $z_0 \in \Omega$; meaning, there is a second order polynomial $P_{z_0}(z)$ such that

$$|u(z) - P_{z_0}(z)| = o(|z - z_0|^2) \quad \text{as } z \to z_0.$$

(see [14]).

The question of uniqueness in the class of *good* solutions is largely an open problem. (See [2], [4], [10], [18] for recent work on the subject). In this paper we describe a general theorem on uniqueness in the case of elliptic operators L with coefficients independent of $t$, more precisely of the form:

$$L = \sum_{h,j=1}^{d} a^{hj}(x)\frac{\partial^2}{\partial x_h \partial x_j} + \sum_{j=1}^{d} b^j(x)\frac{\partial^2}{\partial x_j \partial t} + c(x)\frac{\partial^2}{\partial t^2} \tag{1.5}$$

with coefficients $a^{hj}$, $b^j$, $c \in L^\infty(\Omega)$ that satisfy the uniform ellipticity condition

$$0 < \lambda \le \sum_{h,j=1}^{d} a^{hj}(x)\xi_h\xi_j + \xi_0 \sum_{j=1}^{d} b^j(x)\xi_j + c(x){\xi_0}^2 \le \Lambda \tag{1.6}$$

for every $x \in \Omega$, every $(\xi, \xi_0) \in \mathbf{R}^{d+1}$ with $|\xi|^2 + {\xi_0}^2 = 1$. For such equation we look for *good* solutions $u$ which are uniform limits of sequences $u_n$ satisfying the Dirichlet problem (1.4) where the coefficients of $L_n$ are independent of $t$. We will call such solutions *convolution type good solutions* (see Definition 2.8 for the precise definition).

Define the operator $S$ as

$$S = \sum_{h,j=1}^{d} a^{hj}(x) \frac{\partial^2}{\partial x_h \partial x_j} \tag{1.7}$$

(Observe that (1.6) implies that $S$, as an operator on $\mathbf{R}^d$, is itself uniformly elliptic.)

Our general result states:

**Theorem 1.2** *Let $L$ be as in (1.5) and $S$ as in (1.7) and assume there is uniqueness for the Dirichlet problem in the class of* good *solutions for $S$ and smooth domains $D \subset \mathbf{R}^d$. Then there is uniqueness for the Dirichlet problem (1.3) for $L$ and smooth domains $\Omega \subset \mathbf{R}^{d+1}$ in the class of* convolution type good *solutions.*

The above theorem generalizes the previously obtained result of Luis Escauriaza who proved the same theorem with specialized spatial part $S$ for which uniqueness had previously been established ([6]). Along similar lines but more directed to the question of regularity the interested reader should see the work of Manselli and Papi ([13]), Lamberti and Manselli ([12]) and Arena and Manselli ([1]).

*Remark 1.3.* The hypotheses on $S$ contained in theorem 1.2 is fulfilled if the coefficients of $S$ satisfy the conditions in [18].

The method of proof of the above theorem requires us to prove general results on the uniqueness of *good* solutions to the Dirichlet problem when we perturb a second order homogeneous elliptic operator by adding lower order terms with complex valued coefficients. Briefly, apart from an algebraic condition on the full operator which guarantees a maximum principle for the modulus of a *good* solution to the Dirichlet problem (condition (3.2) in the Appendix), if there is uniqueness in the class of *good* solutions for the homogeneous second order elliptic part, then there is uniqueness for the full operator obtained by adding a gradient term with coefficients in $L^{2d}$ ($d$ = dimension of the ambient space) and a zero order term with coefficients in $L^d$. (See theorem 2.6 and the Appendix.)

One final remark before we begin the proof of our uniqueness theorem. There are other notions of solutions to (1.3). One prominent notion is that of viscosity solution. Recently it has been shown that the notion of *good* solution and that of viscosity solution to (1.3) coincide. This interesting result was proved by Robert Jensen ([9]). In addition to *good* solutions there are other notions of solutions to (1.3) all of which generalize to fully nonlinear equations. A review of these is provided in these proceedings by Crandall, Kocan, Soravia and Swiech [5]. In the linear case these notions coincide with the notion of a *good* solution. This is established in [5] relying on tools developped by R. Jensen in [9].

## 2. Local and global uniqueness

We begin our proof of the main theorem with a discussion in the general setting of the equivalence of the notions of global and local uniqueness in the class of *good* solutions for a general operator $L$ and domain $\Omega$. Of special interest is lemma due to N.V. Krylov ([10]) which generally states that uniqueness in the class of *good* solutions for a given domain $\Omega$ and operator $L$ implies local uniqueness (always in the class of *good* solutions). In this section we also discuss simple cases when the property of uniqueness is preserved after perturbation with a specific zero order term.

We remind the reader that

$$L = \sum_{h,j=1}^{d+1} a^{hj}(z) \frac{\partial^2}{\partial z_h \partial z_j} \quad , \quad z \in \mathbf{R}^{d+1}$$

and that there exist positive constants $\lambda, \Lambda$ such that for all $z$ and $\zeta \in \mathbf{R}^{d+1}$

$$0 < \lambda |\zeta|^2 \leq \sum_{h,j=1}^{d+1} a^{hj}(z) \zeta_h \zeta_j \leq \Lambda |\zeta|^2 \, .$$

We also assume that the matrix $(a_{ij})$ is symmetric.

*Lemma 2.1. If there is uniqueness in the class of* good *solutions to the Dirichlet problem in $\Omega$ with homogeneous boundary data, then there is uniqueness for the full Dirichlet problem in $\Omega$.*

*Proof.* Given $f \in L^{d+1}(\Omega)$ and $g \in C(\partial\Omega)$ we want to show uniqueness in the class of *good* solutions to

$$\begin{cases} Lu = f & \text{in } \Omega \\ u|_{\partial\Omega} = g \end{cases} .$$

The maximum principle implies it is sufficient to take $g$ smooth and defined in all of $\mathbf{R}^{d+1}$. In this case let $v$ be the unique *good* solution to

$$\begin{cases} Lv = f - Lg & \text{in } \Omega \\ v|_{\partial\Omega} = 0 \end{cases}$$

Then the function $u = v + g$ is the (unique) *good* solution to our problem. □

From Lemma 2.1, establishing uniqueness for the full Dirichlet problem is equivalent to showing uniqueness for the Dirichlet problem with homogeneous boundary data, i.e. for the problem

$$\begin{cases} Lu = f & \text{in } \Omega \\ u|_{\partial\Omega} = 0 \end{cases}$$

Throughout the remaining pages this equivalence will be used freely. We will also use the terminology "uniqueness in $\Omega$" to mean uniqueness in the full Dirichlet problem in $\Omega$.

In the next lemma the terminology "local uniqueness in $\Omega$" means for each $z \in \Omega$ there exists $r_z > 0$ such that we have uniqueness in $B_r(z)$ for all $0 < r < r_z$. (Always, uniqueness is taken in the class of *good* solutions).

**Lemma 2.2** *If local uniqueness holds for $L$ in $\Omega$ then uniqueness holds for $L$ in $\Omega$.*

*Proof.* Assume $u_1$ and $u_2$ are *good* solutions to $Lu = f$ in $\Omega$ with $u|_{\partial\Omega} = 0$. If $u_1$ and $u_2$ are different we may assume $u_1 - u_2$ has a global positive maximum $M$ at $z_0$. Local uniqueness implies that there exists $r_{z_0} > 0$ such that for $0 < r < r_{z_0}$

$$M \equiv u_1(z_0) - u_2(z_0) = \int_{\partial B_r(z_0)} \big(u_1(Q) - u_2(Q)\big)\omega_r^{z_0}(dQ)$$

where $\omega_r^{z_0}(dQ)$ is a Borel measure on $\partial B_r(z_0)$ with total mass 1 and positive on open subsets of the boundary. Hence $u_1 - u_2 \equiv M$ on $\partial B_r(z_0)$ and therefore in $B_{r_{z_0}}(z_0)$. Since $\Omega$ is connected this implies $u_1 - u_2 \equiv M$ in $\Omega$ which contradicts the fact that $u_1 - u_2 = 0$ on $\partial\Omega$. □

In the next lemma we will perturb the operator $L$ by subtracting a bounded nonnegative zero order term $c(z)$. The notion of *good* solution for the resulting operator $(L - c)u = Lu - cu$ is the same as for $L$, requiring we construct $L_n - c_n$, $c_n(z)$ to be smooth, nonnegative, bounded in $n$ and in $z$, and converging pointwise almost everywhere to $c(z)$.

**Lemma 2.3.** *Assume uniqueness holds for $L$ in $\Omega$. Then if $c \in L^\infty(D)$, $c \geq 0$ uniqueness holds for $L - c$ in $\Omega$.*

*Proof.* Let $g_{L,\Omega}(z, w)$ denote the Green's function for $L$ in $\Omega$ (unique by hypothesis). Then if $u(z)$ is a *good* solution to

$$\begin{cases} (L - c)u = f & \text{in } \Omega \\ u|_{\partial\Omega} = 0 \end{cases}$$

it satisfies

$$u(z) = -\int_\Omega g_{L,\Omega}(z,w)\, f(w)\, dw \;-\; \int_\Omega g_{L,\Omega}(z,w)\, c(w)\, u(w)\, dw. \tag{2.1}$$

Calling $G_L$ the operator $f \to -\int_\Omega g_{L,\Omega}(z,w)\, f(w)\, dw$ equation (2.1) becomes $u - G_L(cu) = G_L f$ and there exists $\varepsilon = \varepsilon(\lambda, \Lambda, d, \Omega)$ such that for $0 \le c \le \varepsilon$, $u(z)$ is uniquely determined. Here we used the well known fact that

$$\sup_{z\in\Omega} \int_\Omega g_{L,\Omega}(z,w)^{\frac{d+1}{d}}\, dw \;\le C(\lambda, \Lambda, d, \operatorname{diam}\Omega) \quad \text{(see[15])}. \tag{2.2}$$

Now assume $0 \le c \le 2\varepsilon$. Let $c \wedge \varepsilon$ denote the function $\min(c, \varepsilon)$. From what has been proved above there exists exactly one Green's function $g_{L-c\wedge\varepsilon,\Omega}$ for $L - c \wedge \varepsilon$ in $\Omega$ and it is not difficult to show that $g_{L-c\wedge\varepsilon,\Omega} \le g_{L,\Omega}$. We can now write

$$u(z) = -\int_\Omega g_{L-c\wedge\varepsilon,\Omega}(z,w)\,\big(f(w) + (c - c\wedge\varepsilon)(w)\, u(w)\big)\, dw.$$

Since $0 \le c - c\wedge\varepsilon \le \varepsilon$ we once again uniquely determine $u$. Continuing in this manner we uniquely determine the *good* solution $u$ to the problem $(L - c)u = f$ in $\Omega$, $u|_{\partial\Omega} = 0$ for any $c \ge 0$, bounded and measurable. □

The next lemma is a type of converse to Lemma 2.3.

**Lemma 2.4.** *If local uniqueness holds for $L - 1$ in $\Omega$, then it holds for $L$ in $\Omega$. Consequently uniqueness holds for $L$ in $\Omega$.*

*Proof.* Let $B_r$ be a ball of radius $r$ contained in $D$ and $u$ a *good* solution to

$$\begin{cases} Lu = f & \text{in } B_r \\ u|_{\partial B_r} = 0 \end{cases}$$

and let $g_{L-1,B_r}$ be the (unique) Green's function for $L - 1$ in $B_r$. Then $u$ satisfies the integral equation

$$u(z) = -\int_{B_r} g_{L-1,B_r}(z,w)\, f(w)\, dw \;-\; \int_{B_r} g_{L-1,B_r}(z,w)\, c(w)\, u(w)\, dw. \tag{2.3}$$

Since

$$\sup_{z\in B_r} \left(\int_{B_r} g_{L-1,B_r}(z,w)^{\frac{d+1}{d}}\, dw\right)^{\frac{d}{d+1}} \le C(\lambda, \Lambda, d) r^2$$

equation (2.3) uniquely determines $u$ for $r$ sufficiently small. □

We have seen that local uniqueness in $\Omega$ implies uniqueness in $\Omega$. The converse is also true and is due to N.V. Krylov ([10]). We wil need the converse for the operator $L-1$ and when the domain is the special unbounded cylinder $D \times \mathbf{R}$, where $D$ is a smooth bounded domain of $\mathbf{R}^d$. For the sake of completeness and because of the unbounded nature of $\Omega$ we reproduce Krylov's proof in this special setting.

**Lemma 2.5.**(*Krylov*) *Let $z = (x,t) \in \mathbf{R}^{d+1}$. Assume we know uniqueness for $L-1$ in $\Omega \equiv D \times \mathbf{R}$, i.e. for each $f \in C_0^\infty(\Omega)$ there exists a unique* good *solution $u(x,t)$ satisfying*

$$\begin{cases} (L-1)u = f & \text{in } \Omega \\ u(x,t) = 0 & \text{for } x \in \partial D \\ \sup_{x\in D} |u(x,t)| \to 0 & \text{as } t \to \infty \end{cases}$$

*Then local uniqueness holds for $L-1$ in $\Omega$.*

*Proof.* The Green's function $g_{L-1,\Omega}$ corresponding to $L-1$ in $\Omega$ satisfies the inequality

$$g_{L-1,\Omega}(z,w) \leq \int_0^{+\infty} \Gamma(z,w,s)e^{-s}\, ds$$

where $\Gamma(z,w,s)$ is the fundamental solution in $\mathbf{R}^{d+1} \times (0,+\infty)$ corresponding to the operator $L - \partial_s$. From this inequality and the work in [7], we obtain

$$\|g_{L-1,\Omega}(z,\cdot)\|_{L^{\frac{d+1}{d}}(\Omega)} \leq \int_0^\infty \|\Gamma(z,\cdot,s)\|_{L^{\frac{d+1}{d}}(\mathbf{R}^{d+1})} e^{-s}\, ds \leq C(\lambda,\Lambda,d) \tag{2.4}$$

Since we have uniqueness for $L-1$ in $\Omega$ and because of (2.4) the argument in Lemma 2.3 can be used to show uniqueness in $\Omega$ for

$$L^k = L - 1 - k\chi_{\Omega\setminus B},\ \ k \geq 0 \tag{2.5}$$

where $B$ is a ball whose closure is contained in $\Omega$.
Let $u$ be a *good* solution to

$$\begin{cases} (L-1)u = -f & \text{in } B \\ u|_{\partial B} = 0 \end{cases}$$

with $f \in C_0^\infty(B)$. We want to show that $u$ is uniquely determined. We may assume $f \geq 0$. Let $v_k$ be the (unique) *good* solution to $L^k v^k = -f$ in $\Omega$, with $v^k|_{\partial\Omega} = 0$. ($L^k$

is defined by (2.5) and the precise meaning of $v^k|_{\partial\Omega} = 0$ is given in the hypothesis of this lemma). Then $v^k - u$ is a *good* solution in $B = B_r$ of

$$\begin{cases} (L-1)\nu = 0 & \text{in } B \\ \nu|_{\partial B} = v^k|_{\partial B} \end{cases}$$

Hence for $z \in B$,

$$(v^k - u)(z) = \int_{\partial B} v^k(Q)\omega^z_{L-1,B}(dQ)$$

where $\omega^z_{L-1,B}(dQ)$ is a Borel measure on $\partial B$ with total mass $\leq 1$ and positive on open subsets of the boundary. Therefore $u$ will be determined uniquely as $u = \lim_k v^k$ in $B$ if we can prove

$$\lim_{k\to\infty} \sup_{\partial B} v_k = 0.$$

At this point assume the radius of $B$ is small and the center is the origin. Let

$$\psi(z) = A(r^2 - |z|^2),$$

where $A = A(\lambda)$ is chosen so that $L\psi \leq -1$. We also fix a small $\delta > 0$ and can assume for the same $A$ and for $r$ small depending on $A$ that

$$L^k(\psi + 2\delta) \leq -1 \text{ in } B_{(1+\delta)r} \quad \text{and} \quad (\psi + 2\delta)|_{\partial B_{(1+\delta)r}} \geq 0$$

For $z \in B_{(1+\delta)r}$ we have

$$v^k(z) = \int_{B_{(1+\delta)r}} g_{B_{(1+\delta)r},L^k}(z,w)\, f(w)\, dw + \int_{\partial B_{(1+\delta)r}} v^k(Q)\, \omega^z_{B_{(1+\delta)r},L^k}(dQ)$$

where $g_{B_{(1+\delta)r},L^k}$ is the Green's function for $L^k$ in $B_{(1+\delta)r}$ and $\omega^z_{B_{(1+\delta)r},L^k}(dQ)$ is a Borel measure on $\partial B_{(1+\delta)r}$, with total mass $\leq 1$. Now

$$\int_{B_{(1+\delta)r}} g_{B_{(1+\delta)r},L^k}(z,w)\, f(w)\, dw \leq -\sup f \int_{B_{(1+\delta)r}} g_{B_{(1+\delta)r},L^k}(z,w)\, L^k(\psi + 2\delta)(w)\, dw$$

$$\leq \sup f(\psi(z) + 2\delta) = (\sup f)2\delta \quad \text{on } \partial B.$$

Hence to show that $v^k \to 0$ uniformly on $\partial B$ it is sufficent to show this on $\partial B_{(1+\delta)r}$. For each point $z \in \partial B_{(1+\delta)r}$ consider $\overline{B_\rho(z)}$ with $0 < \rho < \frac{\delta r}{2}$. This condition on $\rho$ guaranties $\overline{B_\rho(z)} \subset \Omega \setminus B_r$ and, in $B_\rho(z)$, $v^k$ is a *good* solution to $(L^k - 1 - k)v^k = 0$ with continuous boundary data assigned on $\partial B_\rho(z)$. This means

$$v^k(z) = \int_{\partial B_\rho(z)} v^k(Q)\omega^z_{L-1-k,B_\rho}(dQ)$$

where $\omega^z_{L-1-k,B_\rho}(dQ)$ is a Borel measure obtained as a limit in measure of the sequence of measures

$$\left\{\omega^z_{L^n-1-k,B_\rho}(dQ)\right\}_n$$

and where $L_n = \sum_{h,j=1}^{d+1} a_n^{hj}(z)\frac{\partial^2}{\partial z_h \partial z_j}$ is a sequence of operators with smooth coefficients approximating $L$ (recall definition 1.1). Since $v^k$ are uniformly bounded on $D$ we will complete the proof of the lemma if we can prove

$$\omega^z_{L-1-k,B_\rho(z)}(B_\rho(z)) \leq \frac{c}{k} \tag{2.6}$$

where $c$ depends only on $\rho, \lambda, \Lambda$ and $d$.

To prove this, first of all we may assume the coefficients of $L$ are smooth and let $\Gamma_\rho(z, w, s)$ denote the Green's function corresponding to the parabolic operator $L - \partial_s$ and domain $B_\rho(z) \times (0, \infty)$. Then

$$g_{L-1-k,B_\rho(z)}(z, w) \equiv \int_0^\infty \Gamma_\rho(z, w, s)\, e^{-(k+1)s}\, ds \tag{2.7}$$

is the Green's function for the operator $L - 1 - k$ and domain $B_\rho(z)$. The conormal derivative

$$\frac{\partial}{\partial \nu_Q} g_{L-1-k,B_\rho(z)}(z, Q) \equiv \sum_{h,j=1}^{d+1} \frac{\partial}{\partial w_h} g_{L-1-k,B_\rho(z)}(z, Q)\, a^{hj}(z)\, \frac{Q_j - z_j}{|Q_j - z_j|}$$

for $Q \in \partial B_\rho(z)$, is the Poisson kernel corresponding to the operator $L - 1 - k$ and domain $B_\rho(z)$, and

$$\omega^z_{B_{(1+\delta)r}, L-1-k}\left(B_\rho(z)\right) = \int_{\partial B_\rho(z)} \frac{\partial}{\partial \nu_Q} g_{L-1-k,B_\rho(z)}(z, Q)\, d\sigma(Q)$$

where $d\sigma(Q)$ is a surface measure on $\partial B_\rho(z)$. From formula (2.7) for $g_{L-1-k,B_\rho}$

$$\begin{aligned}\frac{\partial}{\partial \nu_Q} g_{L-1-k,B_\rho(z)}(z, Q) &= \int_0^\infty \frac{\partial}{\partial \nu_Q} \Gamma_\rho(z, Q, s)\, e^{-(k+1)s}\, ds \\ &= \int_0^1 + \int_1^\infty \frac{\partial}{\partial \nu_Q} \Gamma_\rho(z, Q, s)\, e^{-(k+1)s}\, ds.\end{aligned}$$

For $Q \in \partial B_\rho(z)$, $\frac{\partial}{\partial \nu_Q}\Gamma_\rho(\cdot, Q, \cdot)$ is a nonnegative solution to $(L - \partial_s)u = 0$ in $B_\rho(z) \times (0, \infty)$ which is zero initially. The parabolic Harnack inequality ([7]) implies that there exists $C = C(\lambda, \Lambda, \rho, d)$ such that for $0 \leq \tau \leq 1$,

$$\frac{\partial}{\partial \nu_Q}\Gamma_\rho(z,Q,r) \le C\int_2^3 \frac{\partial}{\partial \nu_Q}\Gamma_\rho(z,Q,s)\,ds.$$

Hence

$$\int_0^1 \frac{\partial}{\partial \nu_Q}\Gamma_\rho(z,Q,s)\,e^{-(k+1)s}\,ds \le \frac{\widetilde{C}}{k}\int_2^3 \frac{\partial}{\partial \nu_Q}\Gamma_\rho(z,Q,s)\,e^{-s}\,ds$$

where $\widetilde{C} = \widetilde{C}(\lambda,\Lambda,\rho,d)$. Finally

$$\int_1^\infty \frac{\partial}{\partial \nu_Q}\Gamma_\rho(z,Q,s)\,e^{-(k+1)s}\,ds \le e^{-k}\int_1^\infty \frac{\partial}{\partial \nu_Q}\Gamma_\rho(z,Q,s)\,e^{-s}\,ds.$$

and so we conclude

$$\omega^z_{B_{(1+\delta)r},L-1-k}\left(B_\rho(z)\right) \le \left(\frac{\widetilde{C}}{k}+e^{-k}\right)\omega^z_{B_\rho,L-1}\left(B_\rho(z)\right) \le \left(\frac{\widetilde{C}}{k}+e^{-k}\right)$$

This concludes the proof of inequality (2.6) and of the lemma. □

Only one more result is needed before we combine everything into the proof of the main uniqueness result. The needed theorem is stated here but its proof, which is of independent interest, is given in the Appendix.

Let $D$ be a smooth domain in $R^d$ and let

$$E = \sum_{h,j=1}^d a^{hj}(x)\frac{\partial^2}{\partial x_h \partial x_j} + \sum_{j=1}^d B^j(x)\frac{\partial}{\partial x_j} + C(x) \equiv \tag{2.8}$$

$$\equiv S + \sum_{j=1}^d B^j(x)\frac{\partial}{\partial x_j} + C(x)$$

where $B^j = \beta^j + ib^j$, $C = -c - i\gamma$ are complex valued functions ($\beta^j$, $b^j$, $c$, $\gamma$ real valued) and the coefficients of $S$ satisfy the uniform ellipticity condition

$$0 < \lambda|\xi|^2 \le \sum_{h,j=1}^d a^{hj}(x)\xi_h\xi_j \le \Lambda|\xi|^2$$

for every $x \in D$, every $\xi \in \mathbf{R}^d$.

**Theorem 2.6.** *Let $E$ be as above and assume:*

*(i)* $B = (B_1, ..., B_d) \in L^{2d}(D)$ *and* $C \in L^d(D)$, $c(x) \ge 0$;

*(ii) the coefficients of $E$ satisfy the following condition*

$$c(x) \sum_{h,j=1}^{d} a^{hj}(x)\xi_h\xi_j - \left(\sum_{j=1}^{d} b^j(x)\xi_j\right)^2 \geq 0$$

*for every $x \in D$ and $\xi \in \mathbf{R}^d$;*

*(iii) for $S$ there is local uniqueness in the class of* good *solutions to the full Dirichlet problem.*

*Then there is uniqueness in the class of* good *solutions for the problem*

$$\begin{cases} Eu = f & \text{in } D \\ u|_{\partial D} = 0 \end{cases}$$

*for every $f \in C_0^\infty(D)$.*

*Remark 2.7.* The definition of *good* solution to the operator $L$ given in definition 1.1, carries through in the straightforward way to solutions to $E$, where the convergence of the lower order terms of the approximating operators to those of $E$ is respectively in $L^{2d}(D)$ and in $L^d(D)$.

Before restating and proving our main result let's make precise in which class of solutions we prove uniqueness in Theorem 1.2.

**Definition 2.8.** Let $D \subset \mathbf{R}^d$ and $\Omega \subset D \times \mathbf{R}$ and take $f$ and $g$ smooth. We will say that $u \in C^0(\overline{\Omega})$ is a *convolution type good solution* to the Dirichlet problem

$$\begin{cases} Lu = f & \text{in } \Omega \\ u|_{\partial\Omega} = g \end{cases}$$

if there exists a sequence of operators defined on $D$ of the form (1.5) with coefficients $a_n^{hj}$, $b_n^j$, $c_n$ that satisfy the uniform ellipticity condition (1.6) and such that:

(i) $a_n^{hj}$, $b_n^j$, $c_n$ are smooth in $\overline{\Omega}$ and ;

(ii) $a_n^{hj} \to a^{hj}$, $b_n^j \to b^j$, $c_n \to c$, a. e. in $\Omega$ as $n \to \infty$;

(iii) $L_n u_n = f$ a.e. in $\Omega$, $u_n|_\Omega = g$.

**Main Theorem.** *Let $L$ be as in (1.5) and $S$ as in (1.7) and assume there is uniqueness for the Dirichlet problem in the class of* good *solutions for $S$ and smooth domains*

$D \subset \mathbf{R}^d$. *Then there is uniqueness for the Dirichlet problem (1.3) for $L$ and smooth domains $\Omega \subset \mathbf{R}^{d+1}$ in the class of* convolution type good *solutions.*

*Proof.* Take $\Omega$ a bounded smooth domain in $\mathbf{R}^{d+1}$. To show uniqueness for the full Dirichlet problem

$$\begin{cases} Lu = f & \text{in } \Omega \\ u|_{\partial\Omega} = g \end{cases}$$

with $f \in C_0^\infty(\Omega)$ and $g \in C(\partial\Omega)$ in the class of *convolution type good* solutions we need only show local uniqueness in the same class for the Dirichlet problem with homogeneous boundary data, i.e. $g = 0$. (Lemmas 2.1 and 2.2). We may replace $L$ with $L-1$ (Lemma 2.4) and from Lemma of Krylov it is sufficient to show uniqueness in the class of *convolution type good* solutions $u$ to the following problem

$$\begin{cases} (L-1)u = f & \text{in } \Omega = B \times (-\infty, +\infty) \\ u(x,t) = 0 & \text{for } x \in \partial B \\ \sup_{x \in B} |u(x,t)| \to 0 & \text{as } t \to \infty \end{cases} \tag{2.9}$$

where $B$ is a ball in $\mathbf{R}^d$, $f \in C_0^\infty$.

By translation and dilation, we may assume $B = B_1(0)$. Therefore let $u$ be a *convolution type good* solution to problem (2.9) and let $L_n$ and $u_n$ be the approximating sequences to $L$ and $u$ respectively (see definition 2.8). Consider the auxiliary function $v(t,x) = C(1-|x|^2)e^{-\varepsilon\sqrt{1+t^2}}$. There exist positive constants $C$ and $\varepsilon$, depending on $\lambda$, $\Lambda$ and $\|f\|_\infty$ but independent of $n$, such that

$$L^n v \leq -\|f\|_\infty .$$

The maximum principle now implies that

$$|u_n(t,x)| \leq v(t,x) \quad \text{in } \mathbf{R} \times B \tag{2.10}$$

and, as a consequence, the same estimate holds for $|u(t,x)|$. Moreover, from the interior Schauder estimates, we have that for every $0 < \delta < 1$ and $\bar{t} \in \mathbf{R}$

$$\sup_{|x| \leq 1-\delta, |t-\bar{t}| \leq 1} \sum_{|\alpha| \leq 2} |D^\alpha u_n(t,x)| \leq C_n \left[ \|f\|_{C(K(\delta,\bar{t}))} + \sup_{K(\delta,\bar{t})} |u_n(x,t)| \right] \tag{2.11}$$

where $K(\delta,\bar{t}) = B_{1-\delta/2} \times [\bar{t}-2, \bar{t}+2]$ and $C_n = C(n, \lambda, \Lambda, \delta, \alpha)$.

For $f \in C_0^\infty(\mathbf{R} \times B)$, the above estimate, combined with 2.10, implies that there exist $H$ such that for $|\bar{t}| > H$,

$$\sup_{|x|\leq 1-\delta} \sum_{|\alpha|\leq 2} |D^\alpha u_n(\bar{t}, x)| \leq \tilde{C}_n e^{-\varepsilon\sqrt{1+\bar{t}^2}} \tag{2.12}$$

Let now

$$\widehat{u}_{\nu,n}(x) = \int_{-\infty}^{+\infty} u_n(t,x)e^{-2\pi i\nu t}$$

be the Fourier transform of $u_n$ in the $t$ variable; then, $\widehat{u}_{\nu,n}$ is smooth in $\overline{B}$ and by inequality 2.12 it satisfies

$$L_n^{(\nu)}\widehat{u}_{\nu,n} = \widehat{f}_\nu \text{ in } B, \quad u_{\nu,n}|_{\partial B} = 0,$$

where $L_n^{(\nu)} = \sum_{h,j=1}^{d} a_n^{hj}(x)\frac{\partial^2}{\partial x_h \partial x_j} + \sum_{j=1}^{d} i\nu b_n^j \frac{\partial}{\partial x_j} - \nu^2 c_n$. Moreover (2.10) also implies that for $n$ sufficiently large

$$|\widehat{u}_{\nu,n} - \widehat{u}_\nu| \leq 2\varepsilon + \int_{-H}^{H} |u_n(t,x) - u(t,x)|\ dt$$

and therefore that the uniform convergence of $u_n$ to $u$ in $\mathbf{R} \times B$ implies the uniform convergence of $\widehat{u}_{\nu,n}$ to $\widehat{u}_\nu$ in $B$. From Theorem 2.6 and the hypothesis on $S$, $\widehat{u}_\nu$ is the unique *good* solution to the Dirichlet problem

$$L^{(\nu)}v = \widehat{f}_\nu \text{ in } \Omega, \quad v|_{\partial\Omega} = 0.$$

where $L^\nu = \sum_{h,j=1}^{d} a^{hj}(x)\frac{\partial^2}{\partial x_h \partial x_j} + \sum_{j=1}^{d} i\nu b^j \frac{\partial}{\partial x_j} - \nu^2 c$. Since $\widehat{u}_\nu$ is uniquely determined for each $\nu$, $u$ is uniquely determined and we have proved the thesis. □

## 3. Elliptic operators with complex lower order terms

The goal of this section is to prove theorem 2.6. Therefore we will consider operators $E$ of the form (2.8), i.e.

$$E = \sum_{h,j=1}^{d} a^{hj}(x)\frac{\partial^2}{\partial x_h \partial x_j} + \sum_{j=1}^{d} B^j(x)\frac{\partial}{\partial x_j} + C(x) =$$

where $B^j = \beta^j + ib^j$, $C = -c - i\gamma$ are complex valued functions ($\beta^j$, $b^j$, $c$, $\gamma$ real valued) and satisfying the following conditions:

(i)

$$0 < \lambda|\xi|^2 \leq \sum_{h,j=1}^{d} a^{hj}(x)\xi_h\xi_j \leq \Lambda|\xi|^2 \tag{3.1}$$

for every $x \in D$, every $\xi \in \mathbf{R}^d$ and $a = (a_{ij})$ is symmetric.

(ii) All the coefficients are measurable and $B = (B_1, ..., B_d) \in L^{2d}(D)$ and $C \in L^d(D)$, $c(x) \geq 0$. Fix $N$ satisfying $\|B\|_{L^{2d}(D)} + \|C\|_{L^d(D)} \leq N$.

(iii)

$$c(x) \sum_{h,j=1}^{d} a^{hj}(x)\xi_h\xi_j - \left(\sum_{j=1}^{d} b^j(x)\xi_j\right)^2 \geq 0 \tag{3.2}$$

for every $x \in D$ and $\xi \in \mathbf{R}^d$.

(iv) For $S = \sum_{h,j=1}^{d} a^{hj}(x)\frac{\partial^2}{\partial x_h \partial x_j}$ there is local uniqueness in the class of *good* solutions to the full Dirichlet problem.

We are going to prove:

**Theorem 3.1.** *Let $E$ be as above and assume its coefficients satisfy conditions $(i) - (iv)$. Then there is uniqueness in the class of* good *solutions for the problem*

$$\begin{cases} Eu = f & \text{in } D \\ u|_{\partial D} = 0 \end{cases}$$

*for every $f \in C_0^\infty(D)$.*

To begin with there is some question about the notion of *good* solution: in particular, does one exist and what apriori smoothness does it inherit? These questions arise from the fact that the lower order terms are complex valued and without an obvious maximum principle what apriori estimates can be obtained independent of smoothness assumptions on the coefficients?

We begin our study of the notion of a *good* solution with proving some properties of solutions to equations with complex lower order terms and smooth coefficients. Let $\widetilde{S} = \sum_{h,j=1}^{d} a^{hj}(x)\frac{\partial^2}{\partial x_h \partial x_j} + \sum_{j=1}^{d} \beta^j(x)\frac{\partial}{\partial x_j}$. The first result is a slight refinement of one contained in [1].

**Lemma 3.2.** *Let $u$ be a classical solution of $Eu = f$ in $D$ and $u|_{\partial D} = g$ (with $u$ and $f$ possibly complex valued). Then $|u|$ satisfies*

$$\widetilde{S}|u| \geq -|f| \tag{3.3}$$

*a.e. in the subset of $D$ where $u \neq 0$. As a consequence*

$$\|u\|_{L^\infty(D)} \leq \|g\|_{L^\infty(\partial D)} + K_0 \|f\|_{L^d(D)}, \tag{3.4}$$

*where $K_0 = K_0(\lambda, \Lambda, D, d, N)$*

*Proof.* In order to simplify the notation, in this proof we'll use $v_j$ for $\dfrac{\partial v}{\partial x_j}$ and $v_{jh}$ for $\dfrac{\partial^2 v}{\partial x_j x_h}$. Let $w = |u|$, $u^{(1)} = Re\ u$ and $u^{(2)} = Im\ u$. Moreover let $\zeta_j = u^{(1)} u_j^{(2)} - u^{(2)} u_j^{(1)}$. Then it is easy to check that

$$w^3 Sw = w^2 \left[u^{(1)} Su^{(1)} + u^{(2)} Su^{(2)}\right] + \sum_{j,h=1}^{d} a_{jh}\zeta_j\zeta_h$$

or

$$\widetilde{S}w = \frac{1}{w}\left[u^{(1)} Re\, f + u^{(2)} Im\, f\right] + \frac{1}{w^3}\left[\sum_{j,h=1}^{d} a_{jh}\zeta_j\zeta_h + w^2 \sum_{j=1}^{d} b^j\zeta_j + cw^4\right] \tag{3.5}$$

By condition (3.2) the quantity in the last brackets is nonnegative a.e. Moreover, clearly $\dfrac{u^{(1)}}{w} Re\, f + \dfrac{u^{(2)}}{w} Im\, f \geq -|f|$ and this proves the first part of the Lemma.
To establish the second part, let $v$ be the (unique) solution to the Dirichlet problem $\widetilde{S}v = -|f|$ in $D$, $v|_{\partial D} = |u|\big|_{\partial D}$. By (2.2)

$$\|v\|_{L^\infty(D)} \leq \|u\|_{L^\infty(\partial D)} + K_0 \|f\|_{L^d(D)}\,. \tag{3.6}$$

Therefore the lemma is proved if we prove that $|u| \leq v$ in $D$. The maximum principle implies that $v \geq 0$ in $D$. Also letting $w = |u| - v$, clearly $w \leq 0$ where $u = 0$ and $w|_{\partial D} = 0$. Therefore (3.3) and the maximum principle again imply that $\sup_{\{u \neq O\}} w \leq \sup_{\partial\{u \neq O\}} w$. Since points on $\partial\{u \neq O\}$ are either points on $\partial D$ or points where $u = 0$ we conclude that $w \leq 0$ on all $D$. And the Lemma is proved. □

*Remark 3.3.* Notice that if (3.2) does not hold (3.3) may not hold. As an example, let $d = 2$, $S = \Delta$, $c = \gamma = \beta^j = 0$, $b = \dfrac{-x_j}{|x|}$, $f = 0$. It is easy to see that letting $v(r) = \displaystyle\int_{(3/2)\pi}^{r} e^{it} t^{-1} dt$, $u(x) = v(|x|)$ is a solution to $Lu = 0$ in $2\pi < |x| < 3\pi$ but $|u|$ does not satisfy $\Delta|u| \geq 0$.

*Remark 3.4.* From Lemma 3.2 and general elliptic theory, in the case of smooth coefficients the Dirichlet problem:

$$Eu = f \text{ in } D, \quad u|_{\partial D} = 0$$

has a unique solution for every $f \in \mathcal{C}_0^\infty(D)$.

Moreover the above Lemma implies that there exists a unique Green's function $g_{E,D}$ for $E$ in $D$ (in general complex valued), such that the solution to the above Dirichlet problem is given by the function

$$u(x) = -\int_D g_{E,D}(x,y) f(y)\, dy \quad x \in D.$$

and that for every $x \in D$ and every $f \in L^d(D)$, the following estimate holds

$$\left| \int_D g_{E,D}(x,y) f(y)\, dy \right| \leq \int_\Omega g_{\widetilde{S},D}(x,y) |f(y)|\, dy \leq K_0 \, \|f\|_{L^d(D)} \,. \tag{3.7}$$

where $g_{\widetilde{S},D}$ is the Green's function for $\widetilde{S}$ in $D$. In what follows whenever there is no risk of ambiguity we will not specify the domain $D$, i.e. we'll use $g_E$ instead of $g_{E,D}$.

**Lemma 3.5.** *Let $u$ be a classical solution to $Eu = 0$ in $D$ with $u \in \mathcal{C}^0(\overline{D})$. Then $|u|$ cannot have interior maxima unless it is constant.*

*Proof.* Assume $|u|$ has an interior maximum $M > 0$ at $x_0 \in D$. Then $u \neq 0$ in a ball $\mathcal{B} = \mathcal{B}(x_0)$ and by lemma 3.2 $\widetilde{S}|u| \geq 0$ in $\mathcal{B}$. Since $|u|$ is smooth in $D$, the conclusion now follows from the classical maximum principle which holds for $\widetilde{S}$. □

In the following theorems we'll use the following notation:

$$F = \sum_{j=1}^{d} B^j \frac{\partial}{\partial x_j} + C$$

$$G_S f = \int_D g_S(\cdot, y)\, f(y)\, dy$$

$$G_E f = \int_D g_E(\cdot, y)\, f(y)\, dy;$$

also we'll let $B = (B^1, ..., B^d)$.

**Lemma 3.6.** *Assume the coefficients of $E$ are smooth and satisfy conditions $(i)-(iii)$. Let $u \in \mathcal{C}^0(\overline{D})$ be the classical solution to $Eu = -f$, $u|_{\partial D} = g$, $f$ and $g$ smooth. Then there exist $\alpha = \alpha(d,\, \lambda,\, \Lambda)$ and $\widetilde{K} = \widetilde{K}(d,\, \lambda,\, D,\, \Lambda,\, N)$ such that*

$$|u(x_2) - u(x_1)| \leq \widetilde{K} |x_2 - x_1|^\alpha \left[ \|f\|_{L^d(D)} + \|g\|_{\mathcal{C}^0(\partial D)} \right] . \tag{3.8}$$

*Proof.* The solution $u$ to $Eu = -f$, $u|_{\partial D} = g$ can be written as

$$u = G_S(F(u)) + G_S(f) + w$$

where $w$ satisfies $Sw = 0$ in $D$ and $w|_{\partial D} = g$. It is known that

$$|G_S(f)(x_1) - G_S(x_2)| + |w(x_1) - w(x_2)| \leq K_1 |x_2 - x_1|^\alpha \left[ \|f\|_{L^d(D)} + \|g\|_{\mathcal{C}^0(\partial D)} \right] \tag{3.9}$$

where $0 < K_1 = K_1(d,\ \lambda,\ D,\ \Lambda) < \infty$ and $0 < \alpha = \alpha(d,\ \lambda,\ \Lambda) < 1$.

What remains is to show the apriori Hölder bound for $G_S(F(u))$. We will first show

$$G_S\big(|\nabla u|^2\big) \le K_2\left[\|f\|^2_{L^d(D)} + \|g\|^2_{C^0(\partial D)}\right] \tag{3.10}$$

Observe that, from the ellipticity condition,

$$|\nabla u|^2 = \nabla u \cdot \nabla \bar{u} \le \frac{1}{\lambda}\sum_{i,j=1}^{d} a_{ij}(x)\frac{\partial u}{\partial x_i}\frac{\partial \bar{u}}{\partial x_j} = \frac{1}{2\lambda}\Big(S(u\bar{u}) - u\,S\bar{u} - \bar{u}\,Su\Big).$$

Therefore

$$G_S(|\nabla u|^2) \le \frac{1}{2\lambda}\Big[-|u|^2 - G_S\big(u\,S\bar{u} + \bar{u}\,Su\big)\Big] \le \frac{1}{\lambda}\,G_S\Big(|u|\,|Su|\Big) \le \tag{3.11}$$

$$\le \frac{\|u\|_\infty}{\lambda}\,G_S\Big[|Eu| + |B|\,|\nabla u| + |C|\,|u|\Big],$$

and using the relation between $S$ and $E$, Hölder's inequality and Pucci's result (2.2),

$$\le \frac{\|u\|_\infty}{\lambda}\Big\{G_S(|Eu|) + K_3^{\frac{1}{2}}\,\|B\|_{2d}\left[G_S(|\nabla u|^2)\right]^{1/2} + K_3\ \|C\|_d \cdot \|u\|_\infty\Big\}. \tag{3.12}$$

The second term in the above parenthesis is $\le \dfrac{1}{2}\left[\dfrac{\|u\|_\infty}{\lambda}K_3\,\|B\|_{2d} + \dfrac{\lambda}{\|u\|_\infty}\ G_S(|\nabla u|^2)\right]$,
so that (3.11) and (3.12) lead to

$$G_S(|\nabla u|^2) \le \|u\|^2_\infty\,\frac{K_3}{\lambda}\left[\frac{1}{2\lambda}\ \|B\|^2_{2d} + \|C\|_d\right] + \frac{\|u\|_\infty}{\lambda}\ G_S(|Eu|).$$

From this, using (3.4) and Hölder once more, one easly gets (3.10).

Returning to the Hölder continuity of $G_S\big(F(u)\big)$, let $B_{Mr}$ denote a ball of radius $Mr$.

$$G_S\big(F(u)\big)(x_1) - G_S\big(F(u)\big)(x_2) = \int_D \big(g_S(x_1,y) - g_S(x_2,y)\big)F(u)(y)\ dy = \tag{3.13}$$

$$\int_{D\cap B_{Mr}} + \int_{D\cap B^C_{Mr}} \big(g_S(x_1,y) - g_S(x_2,y)\big)F(u)(y)\ dy.$$

For the first integral above we note that by Hölder's inequality

$$\int_{D\cap B_{Mr}} g_S(x,y)|F(u)(y)|\,dy \le \Big\{\int_{D\cap B_{Mr}} g_S(x,y)|\nabla u|^2\,dy\Big\}^{\frac{1}{2}}\ K_3^{\frac{1}{2}}\ \|B\|_{2d} + K_3\ \|C\|_d\,\|u\|_\infty \tag{3.14}$$

where $K_3 = \sup_x \left( \int_{D \cap B_{Mr}} g_S(x,y)^{\frac{d}{d-1}} dy \right)^{\frac{d-1}{d}}$.
Combining this with (3.10) and using elementary inequalities we obtain

$$G_S(|Fu|) \leq \|u\|_\infty \left\{ \frac{K_3}{\lambda} \left(1 + \frac{1}{\sqrt{2}}\right) \|B\|_{2d}^2 + \frac{3}{2} K_3 \|C\|_d \right\} + \frac{K_3}{2\lambda} \|B\|_{2d}^2 \|G_S(|Eu|))\|_\infty$$

and finally from the results in [7]

$$\int_{D \cap B_{Mr}} g_S(x,z) \big| F(u)(z) \big| \, dz \leq K_4 (Mr)^\alpha \left[ \|f\|_{L^d(D)} + \|g\|_{C^0(\partial D)} \right]. \tag{3.15}$$

Now choose $r = |x_1 - x_2|$ and the center of $B_{Mr}$ equal to $(1/2)(x_1 - x_2)$. Then the Hölder continuity result of Krylov and Safonov gives

$$\int_{D \cap B^C_{Mr}} \big| g_S(x_1,y) - g_S(x_2,y) \big| \, |F(u)(y)| \, dy \leq \widetilde{K} M^{-\alpha} \sup_D G_S(|Fu|) \leq \tag{3.16}$$

$$\leq \widetilde{K} M^{-\alpha} \left[ \|f\|_{L^d(D)} + \|g\|_{C^0(\partial D)} \right].$$

Collecting inequalities (3.13), (3.15) and (3.16) and choosing $M = r^{-1/2}$ we conclude the Hölder continuity bound for $G_S\big(F(u)\big)$. □

Immediately from Lemmas 3.2 and 3.6 we have

*Remark 3.7.* Assume we have a sequence of operators $\{E_n\}$, the coefficients of each operator smooth and satisfying conditions $(i) - (iii)$ with the same $\lambda$, $\Lambda$ and such that

$$\sup_n \left( \|B_n\|_{L^{2d}(D)} + \|C_n\|_{L^d(D)} \right) \leq N < \infty.$$

Let $u_n$ be the solution to $L_n u_n = -f$ in $D$, $u_n|_{\partial D} = g$. Then there exist $\alpha = \alpha(d, \lambda, \Lambda)$ and $\widetilde{K} = \widetilde{K}(d, \lambda, D, \Lambda, N)$ (as in Lemma 3.6) such that

$$\|u_n\|_{L^\infty(D)} \leq \widetilde{K} \left[ \|f\|_{L^d(D)} + \|g\|_{C^0(\partial D)} \right]$$

and

$$\|u_n\|_{C^\alpha(D)} \leq \widetilde{K} \left[ \|f\|_{L^d(D)} + \|g\|_{C^\alpha(\partial D)} \right].$$

As a consequence, for $f \in L^d(D)$ and $g \in C^\alpha(\partial D)$ the sequence $\{u_n\}$ has a subsequence which converges uniformly in $\overline{D}$ and so our notion of *good* solution for an operator $E$

with coefficients satisfying $(i) - (iii)$ is the same as in the case of real valued lower order terms.

The next several Lemmas are designed to show the local uniqueness of *good* solutions under the general assumptions $(i)-(iii)$ on the coefficients. The first Lemma below is a consequence of recent work of L. Caffarelli and it states that *good* solutions associated to the principal part $S = \sum_{h,j=1}^{d} a^{hj}(x)\frac{\partial^2}{\partial x_h \partial x_j}$ of $E$ are differentiable almost everywhere and their gradient is almost everywhere the pointwise limit of the gradients of the approximating sequence.

**Lemma** *(Caffarelli ([3])) Let $u$ be a* good *solution to $Su = f \in L^d(D)$, $u|_{\partial D} = g \in C_0(\partial D)$ $\left(\text{recall } S = \sum_{h,j=1}^{d} a^{hj}(x)\frac{\partial^2}{\partial x_h \partial x_j}\right)$. Assume $u = lim\ u_n$ uniformly in $D$, where $u_n$ satisfies $S_n u_n = f_n \to f \in L^d(\mathcal{D})$, $u_n|_{\partial D} = g$, $\Big(S_n = \sum_{h,j=1}^{d} a_n^{hj}(x)\frac{\partial^2}{\partial x_h \partial x_j}$, $0 < \lambda|\xi|^2 \leq \sum_{h,j=1}^{d} a_n^{hj}(x)\xi_h\xi_j \leq \Lambda|\xi|^2$ and $a_n^{ij} \to a^{ij}$. Then $u$ is differentiable a.e. in $D$,and $\nabla u_n \to \nabla u$ pointwise a.e. in $D$.*

**Lemma 3.9.** *Assume $a^{hj}$ are smooth in $\overline{D}$. Then there exists $\delta > 0$ depending only on $\lambda$, $\Lambda$, $N$, such that if $|D| < \delta$*

$$G_E f = \sum_{k=0}^{\infty} (G_S F)^k G_S f \tag{3.17}$$

*where $G_S F = G_S \circ F$ and $(G_S F)^k = (G_S F) \circ ... \circ (G_S F)$, $k$ times and the series converges uniformly in $\overline{D}$.*

*Proof.* Let $u = -G_E f$; then $Eu = f$ and $u|_{\partial D} = 0$. Therefore, since $Su = -Fu + f$ also $u = G_S(Fu - f)$ or $(I - G_S F)u = -G_S f$. Then we can write formally $u = -\sum_{k=0}^{\infty} (G_S F)^k G_S f$. We need to show that the series converges uniformly.
Let $u_k = (G_S F)^k G_S f$. We'll show that if $|D|$ is small, then $\|u_k\|_\infty \leq \varepsilon^k$, ehere $\varepsilon < 1$. This result is a consequence of the following inequality which follows from (3.14) in Lemma 3.6:

$$\|G_S|Fu|\|_\infty \leq C_0\Big( \|u\|_\infty + \|G_S|Eu|\|_\infty\Big). \tag{3.18}$$

where $C_0 = C(K_3^{\frac{1}{2}} + K_3]$ with $C = C(d, \lambda, \Lambda, N)$ and $K_3 = \sup \left(\int_D g^{\frac{d}{d-1}}\right)^{\frac{d-1}{d}}$. Hence if $|D|$ is small enough the constant $C_0$ can be seen to be $< 1/2$. Taking $u = G_S f$ in (3.18) and using the expression for $E$ we obtain

$$\left\|G_S\big(|F(G_S f)|\big)\right\|_\infty \leq C_0\Big(\left\|G_S f\right\|_\infty + \left\|G_S(|E(G_S f)|)\right\|_\infty\Big) \leq \tag{3.19}$$

$$\leq C_0\Big(2\left\|G_S(|f|)\right\|_\infty + \left\|G_S(|F(G_S f)|)\right\|_\infty\Big)$$

and therefore, being $C_0 < 1/2$,

$$\left\|G_S(|F(G_S f)|)\right\|_\infty \leq C_1\left\|G_S(|f|)\right\|_\infty \tag{3.20}$$

where $C_1 = \dfrac{C_0}{1 - C_0} < 1$. Applying (3.20) to $f = F(G_S f)$ and using (2.2) once more, we obtain

$$\left\|(G_S F)^2(G_S f)\right\|_\infty \leq C_1^2\left\|G_S(|f|)\right\|_\infty \leq C_1^2 C\|f\|_d. \tag{3.21}$$

By recurrence this implies

$$\left\|(G_S F)^k(G_S f)\right\|_\infty \leq C_1^k C\|f\|_d$$

which proves our thesis. □

**Lemma 3.10.** *Assume the coefficients of $E$ satisfy $(i)-(iv)$ and that $D \subset \mathbf{R}^d$ is such that diam $D < M$. Then there exists $\delta = \delta(\lambda, \Lambda, d, N, M) > 0$, such that if $|D| < \delta$ any* good *solution to $Eu = -f \in L^d(D)$ and $u|_{\partial D} = g \in \mathcal{C}(\partial D)$ can be written as*

$$u = \sum_{k=0}^{\infty}(G_S F)^k(G_S f + w_S(g)) \tag{3.22}$$

*(see Lemma 3.9 for notation), where $w_S(f)$ is the* good *solution to the problem $Sw = 0$ in $D$, $w|_{\partial D} = g$. The series in (3.22) converges uniformly in $D$ and as a consequence there is uniqueness in the class of* good *solutions for the above Dirichlet problem.*

*Proof.* First of all remember condition $(iv)$ implies that the Green's function $g_{S,D}$ for $S$ in $D$ is unique and also that $w_S(f)$ is unique. Let

$$S_n = \sum_{h,j=1}^{d} a_n^{hj}(x)\frac{\partial^2}{\partial x_h \partial x_j},$$

where for each $n$ $0 < \lambda|\xi|^2 \leq \sum_{h,j=1}^{d} a_n^{hj}(x)\xi_h\xi_j \leq \Lambda|\xi|^2$ for every $x \in D$, every $\xi \in \mathbf{R}^d$, and $(a_n^{ij})$ is symmetric and $a_n^{hj} \to a^{hj}$ as $n \to \infty$ almost everywhere in $D$. Moreover let $u = \lim_{n\to\infty} u_n$ uniformly in $D$ where $E_n u_n = -f \in L^d(D)$ and $u_n|_{\partial D} = g \in \mathcal{C}(\partial D)$. Recalling Lemma 3.9, $u_n$ can be written as

$$u_n = \sum_{k=0}^{\infty} (G_{S_n} F_n)^k (G_{S_n} f + w_{S_n}(g)). \tag{3.23}$$

Since, from Lemma 3.9 there exists $\delta > 0$ such that if $|D| < \delta$ then $\Big|(G_{S_n} F_n)^k (G_{S_n} f + w_{S_n}(g))\Big| \leq \eta^k$ for some $\eta < 1$, to show that (3.23) holds also for $u$, it is sufficient to show that for each term

$$(G_S F)^k (G_S f + w_S(g)) = \lim_{n\to\infty} (G_{S_n} F_n)^k (G_{S_n} f + w_{S_n}(g)). \tag{3.24}$$

From Lemma 3.8, for $u_{0,n} \equiv G_{S_n} f + w_{S_n}(g)$, $F_n(u_{0,n}$ converges to $F(u_0)$ pointwise almost everywhere in $D$, where $u_0 \equiv G_S f + w_S(g)$. Given $\epsilon > 0$ there exists a measurable set $U_0^\epsilon \subset D$ such that $|D \setminus U_0^\epsilon| \leq \epsilon$ and $F_n(u_{0,n}) \to F(u_0))$ uniformly and boundedly on $U_0^\epsilon$. Let $u_{1,n}^\epsilon = G_{S_n}\big(\chi_{U_0^\epsilon} F_n(u_{0,n})\big)$. Again from Lemma 3.8, there exists a measurable set $U_1^\epsilon \subset D$ such that $|D \setminus U_1^\epsilon| \leq \epsilon$ and $F_n(u_{1,n}^\epsilon) \to F\Big[G_S\big(\chi_{U_0^\epsilon} F(u_0)\big)\Big]$ uniformly and boundedly on $U_1^\epsilon$. Let $u_{2,n}^\epsilon = G_{S_n}\big(\chi_{U_1^\epsilon} F_n(u_{1,n}^\epsilon)\big)$. Proceeding as above we can write $u_{k,n}^\epsilon = G_{S_n}\big(\chi_{\ U_{k-1,n}^\epsilon} F_n(u_{k-1,n}^\epsilon)\big)$ and define

$$u_{k,n} \equiv \Big(G_{S_n} F_n\Big)^k (u_{0,n}) = u_{k,n}^\epsilon + v_{k,n}^\epsilon \tag{3.25}$$

where for $k$ fixed, $|v_{k,n}^\epsilon| = o(1)$ as $\epsilon \to 0$ uniformly in $n$, and $F_n(u_{k-1,n}^\epsilon)$ converges uniformly and boundedly on $U_{k-1,n}^\epsilon$ as $n \to \infty$. Formula (3.25) also gives (3.24), i.e.

$$\lim_{n\to\infty} u_{n,j} = u_j = (G_S F)^j (u_0).$$

Since the series $\sum_{j=0}^{\infty} (G_S F)^j (u_0)$ is uniquely determined through $S$, $f$ and $g$ (condition $(iv)$), the *good* solution $u$ is uniquely determined. □

*Proof (of theorem 2.6.)* We are now ready to prove uniqueness in the class of *good* solutions for operators $E$ satisfying conditions $(i)-(iv)$. So assume $u_1$ and $u_2$ are *good* solutions to the problem $Eu = -f$ in $D$, $u|_{\partial D} = 0$ and assume $|u_1 - u_2|$ has a positive maximum at $x_0 \in D$. From Lemma 3.10 for all sufficiently small balls $B_r$ with center $x_0$ we have for $l = 1, 2$,

$$u_l = \sum_{k=0}^{\infty} \Big(G_{S,B_r} F\Big)^k \Big(G_{S,B_r}(f) + w_{S,B_r}(u_l)\Big)$$

where $G_{S,B_r}(f) = \int_{B_r} g_{S,B_r}(\cdot, y) f(y)\, dy$ and $w_{S,B_r}(u_l)$ is the (unique) solution to the problem $Ew = 0$ in $B_r$, $w|_{\partial B_r} = u_l$.
Therefore

$$u_1 - u_2 = \sum_{j=0}^{\infty} \left(G_{S,B_r} F\right)^j \left(w_{S,B_r}(u_1 - u_2)\right)$$

and from the series representation we identify $u_1 - u_2$ as the unique good solution to the problem $Ew = 0$ in $B_r$, $w|_{\partial B_r} = u_1 - u_2$.
In the proof of Lemma 3.2 we noted that such a $w$ (being a *good* solution) satisfies

$$|w| \leq w_{\widetilde{S},B_r}(|u_1 - u_2|)$$

where $w_{\widetilde{S},B_r}(|u_1 - u_2|)$ is the *good* solution to the Dirichlet problem

$$\widetilde{S}v = \sum_{h,j=1}^{d} a^{hj}(x) \frac{\partial^2 v}{\partial x_h \partial x_j} + \sum_{j=1}^{d} \beta^j(x) \frac{\partial v}{\partial x_j} = 0 \ \ in \ \ B_r, \quad v|_{\partial B_r} = |u_1 - u_2|.$$

This implies that for our $E$ the nonnegative *good* solution $w_{\tilde{S},B_r}(|u_1 - u_2|)$ attains its maximum at $x_0$ the center of $B_r$. Because of Harnack's inequality this in turn implies that $w_{\tilde{S},B_r}(|u_1 - u_2|)$ is identically equal to $\max_{\overline{D}} |u_1 - u_2|$ on $\partial B_r$. Therefore $|u_1 - u_2| = \max_{\overline{D}} |u_1 - u_2|$ in an entire neighborhood of $x_0$. This means $|u_1 - u_2|$ is constant in $D$ and since it is 0 on $\partial D$, $|u_1 - u_2| = 0$ everywhere in $D$. □

## References


[1] Arena, O. and Manselli, P., *A class of elliptic operators in* $\mathbf{R}^3$ *in non-divergence form with measurable coeficients*, Le Matematiche, **XLVIII**, no. 1 (1993), 163-177.

[2] R. Bass, *The Dirichlet Problem for radially homogeneous elliptic operators*, TAMS, **320**, no. 2 (1990), 593-614.

[3] L. Caffarelli, *Elliptic second order equations*, Rendiconti del Seminario Matematico Fisico di Milano, **58** (1988), 253-284.

[4] Cerutti, M.C., Escauriaza, L., and Fabes, E.B., *Uniqueness for the Dirichlet Problem for some Elliptic Equations with Discontinuous Coefficients*, Annali di Matematica pura e applicata (IV), vol. **CLXIII** (1993), 161-180.

[5] M. Crandall, M. Kocun, P. Soravia, A. Swiech *On the equivalence of various weak notions of solutions of elliptic PDES with measurable ingredients*, to appear in this volume.

[6] Escauriaza, L., *Uniqueness for the Dirichlet Problem for time independent elliptic operators*, IMA, vol. **42**, Partial Differential Equations with Minimal Smoothness and Applications, (1992), Springer-Verlag, New York, 115-122.

[7] E. Fabes, D. Strook, *The $L^p$ integrability of Green's functions and fundamental solutions for elliptic and parabolic equations with measurable coefficients*, Duke Math. J., **51** (1984), 977-1016.

[8] Gilbarg, D., and Trudinger, N., *Elliptic Partial Differential Equations of Second Order*, 2nd ed., Springer Verlag, 1983.

[9] R. Jensen, *Uniformly elliptic pdes with bounded measurable coefficients*, to appear in Journal of Fourier Analisys and Applications.

[10] N.V. Krylov, *On one point weak uniqueness for elliptic equations*, Comm. Partial Diff. Equat., **17** (1992), 1759-1784.

[11] N.V. Krylov, M.V. Safonov, *A certain property of solutions of parabolic equations with measurable coefficients*, Izv. Akad. Nauk SSSR Ser. Mat. **44** (1980), 161-175; English translation in Math. USSR Izvestiya, **16** (1981)

[12] L. Lamberti, P. Manselli, *Existence-uniqueness theorems and counterexamples for an axially symmetric elliptic operator*, Bollettino UMI **44** (1983), 431-443.

[13] P. Manselli, G. Papi, *Nonhomogeneous Dirichlet problem for elliptic operators with axially discontinuous coefficients*, Journal of Differential Equations, vol.**55**, no. 3 (1984), 368-384.

[14] S.N. Nadirashvili, *Some differentiability properties of solutions of elliptic equations with measurable coefficients*, Izv. Akad. Nauk SSSR Ser. Mat. **49** (1985), no. 6; English translation in Math. USSR Izvestiya, vol. **27** (1986), no. 3, 601-606.

[15] C. Pucci, *Limitazioni per soluzioni di equazioni ellittiche*, Annali di Matematica pura e applicata, vol. **74** (1966), 15-30.

[16] M.V. Safonov, *Harnack's inequality for elliptic equations and the Hölder property of their solutions*, Zap. Nauchn. Sem. Leningrad, Otdel. Mat. Inst. Steklov (LOMI) **96** (1980), 272-287; English translation in J. Soviet. Math., **21** (1983), no. 5

[17] M.V. Safonov, *Unimprovability of estimates of Hölder constants for solutions of linear elliptic equations with measurable coefficients*, Math. Sbornik, Vol. 132 **174** (1987),no.2; English translation in Math. USSR Sbornik, vol. **60** (1988), no. 1, 269-281.

[18] M.V. Safonov, *On a weak uniqueness for some elliptic equations*, Comm. Partial Diff. Equat., **19**(5 & 6) (1994), 959-1014.

M. Cristina Cerutti
DIPARTIMENTO DI MATEMATICA, POLITECNICO DI MILANO, Piazza Leonardo da Vinci, 32 - 20133 MILANO Italy
e-mail: cricer@ipmma1.polimi.it

Eugene B. Fabes
SCHOOL OF MATHEMATICS, UNIVERSITY OF MINNESOTA, Vincent Hall, 206 Church Street - Minneapolis, MN 55455 USA
e-mail: fabes@math.umn.edu

Paolo Manselli
DIPARTIMENTO DI MATEMATICA, FACOLTÀ DI ARCHITETTURA, UNIVERSITÀ DI FIRENZE, Piazza Brunelleschi - 50100 FIRENZE Italy
e-mail: manselli@cesit1.unifi.it

M.G. CRANDALL[(†)], M. KOCAN[(†)], P. SORAVIA[(‡)], A. ŚWIĘCH

# On the equivalence of various weak notions of solutions of elliptic PDEs with measurable ingredients

## SECTION 0. INTRODUCTION

In this work we outline some aspects of the "$L^p$ theory" of viscosity solutions of scalar fully nonlinear elliptic partial differential equations of the form

$$G(x, u(x), Du(x), D^2u(x)) = 0 \quad \text{in} \quad \Omega, \tag{0.1}$$

where $\Omega$ is a bounded open set in $\mathbb{R}^n$, $Du$ corresponds to the gradient $(u_{x_1}, \ldots, u_{x_n})$ of $u$, $D^2u$ corresponds to the Hessian matrix $(u_{x_i,x_j})$ of second order partial derivatives of $u$, $\mathcal{N} \subset \Omega$ is a Lebesgue null set, and

$$G : (\Omega \setminus \mathcal{N}) \times \mathbb{R} \times \mathbb{R}^n \times \mathcal{S}(n) \to \mathbb{R}.$$

Here $\mathcal{S}(n)$ denotes the set of $n \times n$ real symmetric matrices; $\mathcal{S}(n)$ will be equipped with its usual order, that is $0 \le P$ means $0 \le \langle P\zeta, \zeta\rangle$ for $\zeta \in \mathbb{R}^n$, where $\langle\cdot,\cdot\rangle$ is the Euclidean inner-product, and $X \le Y$ means $0 \le Y - X$. A key point in the discussion is that $G(x, r, p, X)$ depends measurably (and not continuously) on $x$, and therefore the notion of viscosity solutions must be suitably adapted to this situation. Part of our concerns lie in the relations between the various ways this has been done, and we will present some new results in this direction. Other goals of this work are to provide a review of some existing results, and to improve the route to understanding them.

We will rewrite (0.1) in the form

$$F(x, u, Du, D^2u) = f(x) \tag{PDE}$$

by putting $f(x) = -G(x, 0, 0, 0)$, $F(x, r, p, X) = G(x, r, p, X) + f(x)$, so hereafter

$$F(x, 0, 0, 0) = 0.$$

---

(†) Supported in part by National Science Foundation Grant DMS93-02995.
(‡)This work was prepared while visiting University of California at Santa Barbara. Partially supported by Consiglio Nazionale delle Ricerche of Italy.

The structure conditions we will impose herein involve constants $\lambda, \Lambda, \gamma$ with $0 < \lambda \leq \Lambda$, $0 \leq \gamma$ and the Pucci extremal operators given by

$$\mathcal{P}^+(X) = -\lambda \operatorname{tr}\left(X^+\right) + \Lambda \operatorname{tr}\left(X^-\right), \quad \mathcal{P}^-(X) = -\Lambda \operatorname{tr}\left(X^+\right) + \lambda \operatorname{tr}\left(X^-\right), \tag{0.2}$$

where $\operatorname{tr}(X)$ is the trace of $X$, and $X^+$ is the positive part of $X$ (so $\operatorname{tr}\left(X^+\right)$ is the sum of the positive eigenvalues of $X$), etc.

Uniform ellipticity (with constants $\lambda, \Lambda$) and Lipschitz continuity in $p$ (with constant $\gamma$) of $F(x, r, p, X)$ can then be written

$$\mathcal{P}^-(X - Y) - \gamma|p - q| \leq F(x, r, p, X) - F(x, r, q, Y) \leq \mathcal{P}^+(X - Y) + \gamma|p - q| \tag{0.3}$$

for $x \in \Omega$, $r \in \mathbb{R}$, $p, q \in \mathbb{R}^n$ and $X, Y \in \mathcal{S}(n)$. At this juncture we are abusing notation and ignoring the null set $\mathcal{N}$, as we will often hereafter. As regards $r$, we assume that

$$\begin{aligned} &r \to F(x, r, p, X) \quad \text{is nondecreasing and uniformly continuous,} \\ &\text{uniformly for} \quad x \in \Omega \setminus \mathcal{N} \quad \text{and bounded} \quad r, p, X. \end{aligned} \tag{0.4}$$

Important examples of $F$'s satisfying (0.3) include the Pucci extremals (with $\gamma = 0$), the operators

$$\mathcal{P}^+(X) + \gamma|p|, \quad \mathcal{P}^-(X) - \gamma|p|,$$

and the linear case:

$$F(x, r, p, X) = -\operatorname{tr}(A(x)X) + \langle b(x), p\rangle + c(x)r, \tag{0.5}$$

where $A, b, c$ are bounded measurable functions on $\Omega$ with values in $\mathcal{S}(n)$, $\mathbb{R}^n$, $\mathbb{R}$ respectively, and

$$\lambda I \leq A(x) \leq \Lambda I, \quad |b(x)| \leq \gamma, \quad 0 \leq c(x) \quad \text{a.e. in } \Omega. \tag{0.6}$$

Moreover, if each of $F_{\alpha,\beta}(x, r, p, X)$ satisfies (0.3) for $\alpha \in \mathcal{A}$, $\beta \in \mathcal{B}$ ($\mathcal{A}, \mathcal{B}$ being some index sets), then so does

$$F(x, r, p, X) = \sup_{\beta \in \mathcal{B}} \inf_{\alpha \in \mathcal{A}} F_{\alpha,\beta}(x, r, p, X).$$

In Section 1 various notions of subsolutions, supersolutions and solutions of (PDE) are presented. These include the formulations used by Caffarelli, Crandall, Kocan and Święch [3] (which were born in Caffarelli [2]), Jensen [10], as well as other formulations called "strong" (which has the usual meaning), "good" (that is, the limit of strong solutions of approximate problems) and "pointwise a.e.". Section 2 outlines some of the main results of [3]; these are needed in further developments. These results include the deduction of pointwise properties from the $L^p$-viscosity notions used in [3] (e.g., $L^p$-viscosity subsolutions are twice superdifferentiable a.e. and satisfy the relation pointwise a.e), stability properties (a uniform limit of subsolutions is a subsolution of the limit problem, etc.) and the maximum principle of Alexandrov-Bakelman-Pucci. One of the primary tools used is illustrated in the proof of a new existence-uniqueness result. Section 3 examines some further relations between the various notions presented in Section 1. Of particular interest in Section 3 is the clarification of the relationship between pointwise a.e. notions and the others. This observation is new in the generality under discussion, but it is foreshadowed in the classical work of Aleksandrov [1] in the linear case.

In the context of the results mentioned above, the surprising primary result of Jensen [10] states that solutions in other senses given in his paper are also "good solutions" for a class of linear equations. In Section 4 we give an overview of the proof of this very interesting result, incorporating the full range of notions at play in the current work. The uniqueness question for good solutions of Dirichlet problems for linear equations is a compelling unsettled problem in the area, and Jensen's result showed this question is the same if posed for his other notions. Quite interesting special cases of the uniqueness question have been settled, and the latest developments as well as references to the literature are fortunately available in this volume in the contribution of Cerutti, Fabes and Manselli [4]. See also Safonov [13].

Throughout this paper it is assumed that $F$ satisfies the structure conditions (0.3) and (0.4), which we abbreviate to "$F$ has structure":

$$F \quad \text{"has structure" if it satisfies} \quad (0.3), (0.4) \quad \text{and} \quad F(x, 0, 0, 0) \equiv 0.$$

## Section 1. A Plethora of Notions

The theory is replete with notions of solutions. To display this boldly, in this section we introduce the ones which are of concern to us and make some comments about them. The reader need not pursue these in sequence here. To start, it is enough to absorb the first three on the list. Section 1 may be read after this. The remainder can be revisited when these notions arise again in the sequel.

### $L^p$-viscosity Notions

Here is the basic definition used in [3] (and which was born in [2]).

**Definition 1.1.** *Let $F$ have structure, $n < 2p$ (and $1 \leq p$ if $n = 1$) and $f \in L^p_{\mathrm{loc}}(\Omega)$. A function $u \in C(\Omega)$ is an $L^p$-viscosity subsolution (supersolution) of $F = f$ in $\Omega$ if for all $\varphi \in W^{2,p}_{\mathrm{loc}}(\Omega)$, and point $\hat{x} \in \Omega$ at which $u - \varphi$ has a local maximum (respectively, minimum) one has*

$$\begin{cases} \operatorname*{ess\,lim\,inf}_{x \to \hat{x}} (F(x, u(x), D\varphi(x), D^2\varphi(x)) - f(x)) \leq 0 \\ (\operatorname*{ess\,lim\,sup}_{x \to \hat{x}} (F(x, u(x), D\varphi(x), D^2\varphi(x)) - f(x)) \geq 0, \quad \textit{respectively}). \end{cases} \tag{1.1}$$

*Moreover, $u$ is an $L^p$-viscosity solution of $F = f$ in $\Omega$ if it is both an $L^p$-viscosity subsolution and an $L^p$-viscosity supersolution.*

We recall the notion of, e.g, "essential limit inferior" by reformulating the notion of an $L^p$-viscosity subsolution as follows: whenever $\epsilon > 0$, $\mathcal{O} \subset \Omega$ is open, $\varphi \in W^{2,p}_{\mathrm{loc}}(\Omega)$ and

$$F(x, u(x), D\varphi(x), D^2\varphi(x)) - f(x) \geq +\epsilon \quad \text{a.e. in} \quad \mathcal{O} \tag{1.2}$$

then $u - \varphi$ cannot have a local maximum in $\mathcal{O}$. The corresponding restatement of the notion of an $L^p$-viscosity supersolution is evident.

The range $n < 2p$ used in Definition 1.1 was chosen because this is the range in which $W^{2,p}_{\mathrm{loc}}(\Omega)$ functions are continuous and twice differentiable a.e. (see, e.g., Appendix C of [3] regarding this classical result). However, it is not the range in which there is a theory. Fixing the ellipticity $\mathcal{E}$ $(=\Lambda/\lambda)$, it turns out that for $n > 1$ there is an $\epsilon > 0$ such that the range $n - \epsilon < p$ is appropriate for the equation $F = f$. In this presentation we will restrict ourselves to $n \leq p$ (and $p > 1$ if $n = 1$) for simplicity. We leave it to the reader to consult [3] for the general case, which relies on Escauriaza's extension to $n - \epsilon < p$ ([6]) of the fundamental arguments of Caffarelli [2].

A basic fact concerning the notion of a solution in Definition 1.1 is that the uniqueness of solutions of Dirichlet problems with this notion of solution of the partial differential equations remains a major open problem in the field.

## $C$-viscosity Notions

The notions defined in Definition 1.1 depend on the choice of a class of "test functions" $\varphi$ – this class was taken to be $W^{2,p}_{\mathrm{loc}}(\Omega)$ in the definition, but we obtain another notion whenever we vary the class. For example, we might take the "classical choice" $C^2(\Omega)$ and thus define $C$-viscosity subsolutions, supersolutions, etc. If $F(x, r, p, X)$, $f(x)$ are continuous in $x$, and $\varphi \in C^2$, the limits in (1.1) may be replaced by pointwise evaluation, and we have the standard notions of [5]. More generally, owing to the continuity in $(r,p,X)$ always assumed of $F$, if $\varphi \in C^2$, (1.1) is equivalent to

$$\begin{cases} \operatorname{ess}\liminf_{x\to\hat{x}}(F(x,u(\hat{x}),D\varphi(\hat{x}),D^2\varphi(\hat{x}))-f(x))\le 0 \\ (\operatorname{ess}\limsup_{x\to\hat{x}}(F(x,u(\hat{x}),D\varphi(\hat{x}),D^2\varphi(\hat{x}))-f(x))\ge 0). \end{cases} \tag{1.3}$$

A serious defect of the $C$-viscosity notions is made clear by the following example from [3]:

**Example 1.2.** Let $A\subset[-1,1]$ be a measurable set such that for all nontrivial subintervals $I\subset[-1,1]$, we have $|A\cap I|$ and $|A^c\cap I|>0$, where $A^c=[-1,1]\setminus A$ and $|E|$ is the Lebesgue measure of $E$. Let $\chi_A$ be the characteristic function of $A$ and $f(x)=\chi_A-\chi_{A^c}$. Since

$$\operatorname{ess}\liminf_{y\to x} f(y)=-1 \quad \text{and} \quad \operatorname{ess}\limsup_{y\to x} f(y)=1$$

for $x\in(-1,1)$, one quickly verifies from the above that every $v\in C^\infty([-1,1])$ such that $|v''|\le 1$ is a $C$-viscosity solution of $-u''=f$ in $(-1,1)$. Hence there are infinitely many $C$-viscosity solutions of the Dirichlet problem $-u''=f$ in $(-1,1)$, $u(-1)=u(1)=0$.

There is, however, only one $L^\infty$-viscosity solution of the above Dirichlet problem, which is in fact an $L^\infty$-strong solution in the sense of the next definition.

### $L^p$-strong Notions

The concept of strong solutions, etc., has its usual meaning.

**Definition 1.3.** *Let $n\le p\le\infty$ and $f\in L^p_{\text{loc}}(\Omega)$. An $L^p$-strong subsolution (respectively, an $L^p$-strong supersolution, solution) of $F=f$ in $\Omega$ is a function $\psi\in W^{2,p}_{\text{loc}}(\Omega)$ such that $F(x,\psi(x),D\psi(x),D^2\psi(x))\le f(x)$ (respectively, $\ge f(x)$, $=f(x)$) a.e. in $\Omega$.*

It is obvious that the Dirichlet problem of Example 1.2 has a unique solution $u\in C([-1,1])$ which is an $L^\infty$-strong solution of $-u''=f$ on $(-1,1)$. We will learn, in Section 2, that this strong solution is the unique $L^p$-viscosity solution for $1\le p\le\infty$.

The uniqueness of strong solutions of Dirichlet problems is classical and easy to establish, but in general strong solutions do not exist.

## The Pointwise a.e. Notions

We recall the notation: $(p, X) \in \mathcal{D}^{2,+}u(\hat{x})$ if

$$u(x) \leq u(\hat{x}) + \langle p, x - \hat{x}\rangle + \tfrac{1}{2}\langle X(x - \hat{x}), x - \hat{x}\rangle + o(|x - \hat{x}|^2) \tag{1.4}$$

as $x \to \hat{x} \in \Omega$. Moreover, we recall that $(p, X) \in \mathcal{D}^{2,+}u(\hat{x})$ exactly when there exists $\varphi \in C^2(\Omega)$ such that $u - \varphi$ has a strict maximum at $\hat{x}$ and $D\varphi(\hat{x}) = p$, $X = D^2\varphi(\hat{x})$. Moreover $(p, X) \in \mathcal{D}^{2,-}u(\hat{x})$ if (1.4) holds when $\leq$ is replaced by $\geq$, etc.

The pointwise a.e. notions are defined as in the $L^p$-strong notions, but do not require regularity of the functions involved. We begin restricting ourselves to the case of subsolutions, the cases of supersolutions and solutions are then evident.

**Definition 1.4.** *Let $F$ have structure, $n \leq p$, and $f \in L^p(\Omega)$. Then $u \in C(\Omega)$ is a pointwise a.e. viscosity solution of $F \leq f$ in $\Omega$ if and only if*

$$\begin{cases} \text{for almost all} \quad x \in \Omega \quad \text{we have} \quad F(x, u(x), p, X) \leq f(x) \\ \text{whenever} \quad (p, X) \in \mathcal{D}^{2,+}u(x) \end{cases} \tag{1.5}$$

*or, equivalently,*

$$\begin{cases} \text{for almost all} \quad x \in \Omega \quad \text{we have} \quad F(x, u(x), D\varphi(x), D^2\varphi(x)) \leq f(x) \\ \text{whenever} \quad \varphi \in C^2(\Omega) \quad \text{and} \quad u - \varphi \quad \text{has a strict maximum at} \quad x. \end{cases} \tag{1.6}$$

The pointwise a.e. notions will prove useful only under further restrictions. For example, the functions $v(x) = -|x|$ and $w(x) = |x|$ are both pointwise a.e. solutions of $-u''(x) = 0$ in $\mathbb{R}$. However, neither is a solution in the other senses above, while $w$ is in fact an $L^1$-viscosity subsolution in the sense of Definition 1.1.

## Jensen's Notions

We turn to some of the notions of Jensen [10], who proposes a number of them and shows them all to be equivalent. In Jensen's context, he is discussing the purely second order linear case

$$F(x, D^2u) = f(x), \quad \text{where} \quad F(x, X) = -\text{tr}\,(A(x)X) \tag{LE}$$

and $\lambda I \leq A(x) \leq \Lambda I$ a.e. Moreover, he takes $f \in L^\infty(\Omega)$. He tests with $C^2$ functions, and we made the distinct point that we could not take $\varphi \in C^2$ in Definition 1.1. The example establishing this was of Jensen's type with $n = 1$. We eventually resolve these issues in Section 3.

For $r > 0$ and $x \in \mathbb{R}^n$, $B_r(x)$ will denote the closed ball of radius $r$ centered at $x$. Let

$$\fint_E f(x)\,dx = \frac{1}{\text{meas}(E)} \int_E f(x)\,dx.$$

Jensen's weakest notion proposes:

**Definition 1.5.** *$u \in C(\Omega)$ is a $\gamma$-subsolution of (LE) if for every $\varphi \in C^2(\Omega)$ and maximum $\hat{x}$ of $u - \varphi$ and $\eta > 0$ one has*

$$\limsup_{r \downarrow 0} \fint_{B_r(\hat{x})} \left(\text{tr}\left(A(x)(D^2\varphi(x) + \eta I)\right) + f(x)\right)^+ dx > 0. \tag{1.7}$$

We apologize to the reader for the conflict between our structure constant $\gamma$ and the usage "$\gamma$-subsolution", which has nothing to do with this constant.

Jensen's strongest notion proposes:

**Definition 1.6.** *$u \in C(\Omega)$ is a $\beta$-subsolution of (LE) if for every $\varphi \in C^2(\Omega)$ and maximum $\hat{x}$ of $u - \varphi$ and $\eta > 0$ one has*

$$\fint_{B_r(\hat{x})} \left(\text{tr}\left(A(x)(D^2\varphi(x) + \eta I)\right) + f(x)\right)^+ dx \geq C_J \tag{1.8}$$

*for small $r > 0$ for some constant $C_J > 0$ depending only on $\eta, n, \lambda, \Lambda$ and $L^\infty$ bounds on $D^2\varphi$ and $f$.*

A suitable explicit and complicated choice of the constant $C_J$ is given in Jensen [10], but this is not really relevant for us. Our awkward formulation will be made clear in Section 3.

### The Notion of a Good Solution

The notion of "good solutions" is the "limit of strong solutions" of approximate problems. We take $p = n$ and explain a "good" sequence of approximations $F_m = f_m$ of $F = f$ – we ask that $F_m$ have structure and $F_m \to F, f_m \to f$ in the following sense:

For every $\psi \in W^{2,n}_{\text{loc}}(\Omega)$, $F_m(x, \psi, D\psi, D^2\psi) - f_m \to F(x, \psi, D\psi, D^2\psi) - f$ in $L^n_{\text{loc}}(\Omega)$.

One says that $u$ is a good solution of $F = f$ on $\Omega$ if there is a good sequence of approximations $F_m = f_m$ of $F = f$ and $L^n$-strong solutions $u_m$ of $F_m = f_m$ on $\Omega$

such that $u_m \to u$ uniformly on compact subsets of $\Omega$. If $F$ is linear, we require each $F_m$ to be linear.

## Section 2. Main Results With the $L^p$-viscosity Notions and Some Relations

We are primarily concerned here with the $L^p$-viscosity notions. The results presented below are largely from [3], where one finds further references to their origins. The order in which we present these results is not the order in which they are proved, but is chosen for ease of understanding.

We begin by noting some relations among the (apparently $p$-dependent) $L^p$-strong, $L^p$-viscosity and $C$-viscosity notions. The next result states, roughly, that if two of them are appropriate for the same equation, then they coincide for that equation.

**Theorem 2.1.** *Let $F$ have structure, $n \le p < \infty$ and $f \in L^p_{\text{loc}}(\Omega)$.*

*(i) If $u$ is an $L^p$-strong solution of $F \le f$ in $\Omega$, then $u$ is an $L^p$-viscosity solution of $F \le f$ in $\Omega$.*

*(ii) If $F, f$ are continuous in $x$, then $u$ is a $C$-viscosity solution of $F \le f$ in $\Omega$ if and only if $u$ is an $L^p$-viscosity solution of $F \le f$ in $\Omega$.*

*(iii) If $n \le q < p$, then $u$ is an $L^q$-viscosity solution of $F \le f$ in $\Omega$ if and only if $u$ is an $L^p$-viscosity solution of $F \le f$ in $\Omega$.*

*(iv) Statements (i)-(iii) remain true with $F \le f$ replaced by either $F \ge f$ or $F = f$.*

**Remark 2.2.** According to Theorem 2.1 (iii), we may as well restrict our attention to $L^n$-viscosity notions in most statements, and we will do so. The assertion (i) follows from Bony's Maximum Principle [12] as well as from results below, (ii) is Proposition 2.9 in [3] and (iii) is announced here for the first time. The awkward proof of (ii) in [3] is due to the sequencing of results in that presentation; we indicate a better proof by establishing (iii) at the end of this section.

We next state the version of the Aleksandrov-Bakelman-Pucci maximum principle proved in [3]. To appreciate it, note that by (0.3) with $q = 0, Y = 0$ and (0.4), an $L^n$-viscosity solution of $F \le f$ (that is, an $L^n$-viscosity subsolution of $F = f$) is also an $L^n$-viscosity solution of $\mathcal{P}^-(D^2u) - \gamma|Du| \le f(x)$ on $\{u > 0\}$, etc.

In the statement we use the following notations. If $w: \Omega \to \mathbb{R}$, then

$$\Gamma^+(w, \Omega) = \{x \in \Omega : \exists p \in \mathbb{R}^n \text{ such that } w(y) \le w(x) + \langle p, y - x\rangle \text{ for } y \in \Omega\}.$$

$\Gamma^+(w,\Omega)$ is called the "upper contact set" of $w$. Moreover, if $\kappa > 0$, then

$$\Gamma^+_\kappa(w,\Omega) = \{x \in \Omega : \exists p \in \mathbb{R}^n, |p| \le \kappa, \text{ such that } w(y) \le w(x) + \langle p, y-x\rangle \text{ for } y \in \Omega\}.$$

**Proposition 2.3.** *Let $F$ have structure and $f \in L^n(\Omega)$. There exists a constant $C(\lambda,\gamma,n)$ depending only on the quantities indicated, such that if $u \in C(\overline{\Omega})$ is an $L^n$-viscosity solution of*

$$\mathcal{P}^-(D^2u) - \gamma|Du| \le f(x) \quad on \quad \{0 < u\},$$

*then for $0 < \kappa \le (\sup_\Omega u - \sup_{\partial\Omega} u^+)/\operatorname{diam}(\Omega)$*

$$\kappa \le C(\lambda,\gamma,n)\|f^+\|_{L^n(\Gamma^+_\kappa(u^+,\Omega))}.$$

*In particular,*

$$\sup_\Omega u \le \sup_{\partial\Omega} u^+ + \operatorname{diam}(\Omega)\cdot C(\lambda,\gamma,n)\cdot \|f^+\|_{L^n(\Gamma^+(u^+,\Omega))}.$$

*Similarly, if $u$ is an $L^n$-viscosity solution of*

$$f(x) \le \mathcal{P}^+(D^2u) + \gamma|Du| \quad on \quad \{u < 0\},$$

*then*

$$\sup_\Omega u^- \le \sup_{\partial\Omega} u^- + \operatorname{diam}(\Omega)\cdot C(\lambda,\gamma,n)\cdot \|f^-\|_{L^n(\Gamma^+(u^-))}.$$

The point of this result is its technical generality. It is a more precise version of Proposition 3.3 in [3], whose proof establishes all that is claimed above.

**Remark 2.4.** Hereafter, noting that $u$ is a subsolution (according to any notion in use) of $F = f$ if and only if $v = -u$ is a supersolution of $G = -f$ where $G(x,r,p,X) = -F(x,-r,-p,-X)$ and $G$ has structure iff $F$ has structure, we reemphasize that we will formulate most statements for subsolutions only. For example, the supersolution case in Proposition 2.3 follows from the subsolution case, since $\mathcal{P}^+(X) + \gamma|p| = -(\mathcal{P}^-(-X) - \gamma|-p|)$.

The tool provided by Proposition 2.3 is used in the proofs of all other results. An example of the use of Proposition 2.3 is the following: Suppose $u, v \in C(\overline{\Omega})$, $u$ is an

$L^n$-viscosity solution of $F \le f$ and $v$ is an $L^n$-strong solution of $F \ge f$ in $\Omega$. Then, formally, on $\{u - v > 0\} = \{u > v\}$ we have

$$\begin{aligned}\mathcal{P}^-(D^2u - D^2v) - \gamma|D(u-v)| &\le F(x, v, Du, D^2u) - F(x, v, Dv, D^2v)\\ &\le F(x, u, Du, D^2u) - F(x, v, Dv, D^2v) \le f - f = 0.\end{aligned}$$

The first inequality is from (0.3) and the second from the monotonicity in (0.4) and $u > v$ on the set in question. Using that $u$ is an $L^n$-viscosity subsolution and $v$ is an $L^n$-strong supersolution, one easily checks that the extreme inequality is correct in the $L^n$-viscosity sense. (The reader is advised that it would be a major advance in the theory if this calculation was shown to be correct when $v$ was merely an $L^n$-viscosity supersolution as well.) Applying Proposition 2.3 we find $u - v \le \max_{\partial\Omega}(u-v)^+$ in $\Omega$, so if $u \le v$ on $\partial\Omega$, then $u \le v$ in $\Omega$. It follows that an $L^n$-strong solution of $F = f$ in $\Omega$ is the unique $L^n$-viscosity solution assuming the same boundary values; $L^n$-strong solutions of Dirichlet problems are unique among $L^n$-viscosity solutions.

We turn to pointwise properties of $L^p$-viscosity subsolutions, etc. We say $u$ is twice superdifferentiable at $\hat{x}$ if the set $\mathcal{D}^{2,+}u(\hat{x})$ of (1.4) is nonempty. $\mathcal{D}^{2,-}u(\hat{x})$ being nonempty corresponds to "twice subdifferentiablity". We have $(p, X) \in \mathcal{D}^{2,+}u(\hat{x}) \cap \mathcal{D}^{2,-}u(\hat{x})$ if and only if

$$u(x) = u(\hat{x}) + \langle p, x - \hat{x}\rangle + \tfrac{1}{2}\langle X(x-\hat{x}), x - \hat{x}\rangle + o(|x - \hat{x}|^2).$$

This last relation is the definition of "$u$ is twice differentiable at $\hat{x}$ and $p = Du(\hat{x})$, $X = D^2u(\hat{x})$".

**Theorem 2.5.** *Let $F$ have structure and $f \in L^n(\Omega)$. Then there is a null set $\mathcal{N}' \subset \Omega$ such that if $u$ is an $L^n$-viscosity solution of $F \le f$ in $\Omega$ and $x \in \Omega \setminus \mathcal{N}'$ and $(p, X) \in \mathcal{D}^{2,+}u(x)$, then*

$$F(x, u(x), p, X) \le f(x).$$

**Theorem 2.6.** *Let $f \in L^n_{\text{loc}}(\Omega)$ and $u$ be an $L^n$-viscosity solution of $\mathcal{P}^-(D^2u) - \gamma|Du| \le f$ in $\Omega$. Then $u$ is twice superdifferentiable a.e. in $\Omega$.*

**Corollary 2.7.** *If $F$ has structure, $f \in L^n_{\text{loc}}(\Omega)$ and $u$ is an $L^n$-viscosity solution of $F = f$, then $u$ is twice differentiable a.e. and the pointwise derivatives satisfy the equation a.e.*

Theorem 2.5, Theorem 2.6, Corollary 2.7 correspond to information in Propositions 3.4 and 3.5 of [3].

The next theorem states roughly that uniform limits of solutions are solutions, a property we refer to as "stability".

**Theorem 2.8.** *Let $F_m(x,r,p,X), F(x,r,p,X)$ have structure, let $f, f_m \in L^n(\Omega)$ and let $u_m \in C(\Omega)$ be $L^n$-viscosity subsolutions of*

$$F_m(x, u_m, Du_m, D^2u_m) = f_m \quad in\ \Omega \tag{2.1}$$

*for $m = 1, 2, \ldots$. Assume that for $\varphi \in W^{2n}_{\text{loc}}(\Omega)$, if*

$$g_m(x) = F_m(x, u_m(x), D\varphi(x), D^2\varphi(x)) - f_m(x)$$

*and*

$$g(x) = F(x, u(x), D\varphi(x), D^2\varphi(x)) - f(x),$$

*then*

$$(g - g_m)^+ \to 0 \quad \text{in} \quad L^n_{\text{loc}}(\Omega) \tag{2.2}$$

*and $u_m \to u$ locally uniformly as $m \to \infty$. Then $u$ is an $L^n$-viscosity subsolution of $F = f$.*

See Theorem 3.8 in [3].

The following Lemma is an essential tool in the proofs of the above results. We call it a "lemma", but its proof is based on the regularity results of Caffarelli [2] as extended by Escauriaza [6], and these results are deep.

**Lemma 2.9.** *Let $f \in L^p(\Omega)$, $n \le p < \infty$, $\psi \in C(\partial\Omega)$, $\gamma \ge 0$ and $\Omega' \subset\subset \Omega$. Then there exists constants*

$$C_0 = C(n, \gamma, p, \lambda, \Lambda, \text{diam}\,(\Omega), \text{dist}(\Omega', \partial\Omega)), \quad C_1 = C_1(\gamma, \lambda, \Lambda, n, \text{diam}\,(\Omega))$$

*depending only on the quantities indicated, and a strong solution $u \in C(\overline{\Omega}) \cap W^{2,p}_{\text{loc}}(\Omega)$ of*

$$\mathcal{P}^+(D^2u) + \gamma|Du| \le f \tag{2.3}$$

*in $\Omega$ such that $u = \psi$ on $\partial\Omega$. Moreover, $u$ satisfies*

$$\|u\|_{W^{2,p}(\Omega')} \le C_0 \left( \|\psi\|_{L^\infty(\partial\Omega)} + \|f\|_{L^p(\Omega)} \right) \tag{2.4}$$

*and*

$$\|u\|_{L^\infty(\Omega)} \le \|\psi\|_{L^\infty(\partial\Omega)} + (\text{diam}\,(\Omega))^{2-\frac{n}{p}} C_1 \|f\|_{L^p(\Omega)}. \tag{2.5}$$

*Finally, if* $0 \leq \psi, f$ $(\psi, f \leq 0)$*, then* $0 \leq u$ *(respectively,* $u \leq 0$*).*

See Lemma 3.1 in [3].

We illustrate the use of this tool in the following proof.

**Proposition 2.10.** *Suppose that* $F(p, X)$ *has structure,* $\Omega$ *has the exterior sphere property and* $f \in L^n(\Omega)$*. Then the Dirichlet problem*

$$F(Du, D^2u) = f(x) \quad in \quad \Omega, \qquad u = 0 \quad on \quad \partial\Omega \tag{2.6}$$

*has a unique solution* $u \in C(\overline{\Omega})$ *which is an* $L^n$*-viscosity solution of* $F = f$ *in* $\Omega$*.*

*Proof.* Let $f_j \in C(\overline{\Omega})$ converge to $f$ in $L^n(\Omega)$. By the standard methods exposed in [5] (and see [8]), the problem

$$F(Du_j, D^2u_j) = f_j(x) \quad \text{in} \quad \Omega, \quad u_j = 0 \quad \text{on} \quad \partial\Omega \tag{2.7}$$

has a unique $C$-viscosity solution $u_j \in C(\overline{\Omega})$, which by Theorem 2.1 is also an $L^n$-viscosity solution. Let $\psi_{j,k} \in C(\overline{\Omega}) \cap W^{2,n}_{\text{loc}}(\Omega)$ be the solutions of

$$\mathcal{P}^-(D^2\psi_{j,k}) - \gamma|D\psi_{j,k}| \geq (f_j - f_k) \quad \text{in} \quad \Omega, \quad \psi_{j,k} = 0 \quad \text{on} \quad \partial\Omega$$

provided by Lemma 2.9. Then, in the $L^n$-viscosity sense and consequently in the $C$-viscosity sense,

$$\begin{aligned} F(D(u_k + \psi_{j,k}), D^2(u_k + \psi_{j,k})) &\geq \mathcal{P}^-(D^2\psi_{j,k}) - \gamma|D\psi_{j,k}| + F(Du_k, D^2u_k) \\ &\geq (f_j - f_k) + f_k = f_j = F(Du_j, D^2u_j) \end{aligned}$$

and so by the comparison theory of $C$-viscosity solutions of (2.7) $u_j \leq u_k + \psi_{j,k}$, which implies, via Proposition 2.3,

$$(u_j - u_k) \leq \psi_{j,k} \leq C\|f_j - f_k\|_{L^n(\Omega)} \to 0.$$

We conclude that $u_j \to u$ for some $u$, which is an $L^n$-viscosity solution of $F = f$ by Theorem 2.8. The uniqueness is shown by the same method; we will show that any $L^n$-viscosity solution $u$ of (2.6) is the limit of the $u_j$ as in (2.7). Use Lemma 2.9 to find $\psi_j \in C(\overline{\Omega}) \cap W^{2,n}_{\text{loc}}(\Omega)$ such that

$$\mathcal{P}^-(D^2\psi_j) - \gamma|D\psi_j| \geq (f_j - f) \quad \text{in} \quad \Omega, \quad \psi_j = 0 \quad \text{on} \quad \partial\Omega.$$

Then

$$\begin{aligned} F(D(u + \psi_j), D^2(u + \psi_j)) &\geq \mathcal{P}^-(D^2\psi_j) - \gamma|D\psi_j| + F(Du, D^2u) \\ &\geq (f_j - f) + f = f_j = F(Du_j, D^2u_j) \end{aligned}$$

in the $L^n$-viscosity sense and therefore in the $C$-viscosity sense, and by comparison $u + \psi_j \geq u_j$ and then $u_j - u \leq \psi_j \leq C\|f_j - f\|_{L^n(\Omega)} \to 0$. By symmetric arguments also $(u - u_j)^+ \to 0$. □

We think this result is proved here for the first time; its proof parallels that of Example 3.12 in [3]. It remains a major open problem whether or not the solution obtained in Proposition 2.10 above is an $L^n$-strong solution.

We finish this section with the proof of part (iii) of Theorem 2.1.

*Proof of Theorem 2.1 (iii).* It is obvious that if $p \geq q$ and $f \in L^p_{\text{loc}}(\Omega)$ then $L^q$-viscosity solutions of $F \leq f$ in $\Omega$ are $L^p$-viscosity solutions. So suppose that $u$ is an $L^p$-viscosity subsolution and not an $L^q$-viscosity subsolution of $F \leq f$. Then one can find $B_r(\hat{x}) \subset\subset \Omega$ and $\psi \in W^{2,q}$ such that $u - \psi$ has a strict local maximum at $\hat{x}$ and

$$F(x, u, D\psi, D^2\psi) \geq f \quad \text{a.e. in} \quad B_r(\hat{x}).$$

We may assume that

$$u(\hat{x}) - \psi(\hat{x}) > \delta > 0 \quad \text{and} \quad u - \psi < -\delta \quad \text{on} \quad \partial B_r(\hat{x}). \tag{2.8}$$

Let $\varphi_m \in C^2(\Omega)$ converge to $\psi$ in $W^{2,q}$ (and consequently in $W^{2,n}$). Then $w_m = u - \varphi_m$ is an $L^p$-viscosity subsolution of

$$\mathcal{P}^-(D^2 w_m) - \gamma|Dw_m| \leq f(x) - F(x, u, D\varphi_m, D^2\varphi_m) \equiv g_m(x) \quad \text{in} \quad B_r(\hat{x}).$$

Note that $g_m \in L^p(B_r(\hat{x}))$ and $g_m(x) \leq F(x, u, D\psi, D^2\psi) - F(x, u, D\varphi_m, D^2\varphi_m)$ shows that

$$\|g_m^+\|_{L^n(B_r(\hat{x}))} \to 0 \quad \text{as} \quad m \to \infty. \tag{2.9}$$

Using a version of the maximum principle Proposition 2.3, where $u$ is an $L^p$-viscosity subsolution and $f \in L^p$, $p \geq n$, (which is not stated in [3] but follows easily from the proof of Proposition 3.3 there) we obtain

$$\sup_{B_r(\hat{x})} w_m \leq \sup_{\partial B_r(\hat{x})} w_m^+ + C\|g_m^+\|_{L^n(B_r(\hat{x}))},$$

which together with (2.8) and (2.9) yields a contradiction. □

## Section 3. Relations Between $L^p$-viscosity, Pointwise a.e. and Jensen's Notions

Suppose that $F$ has structure, $f \in L^n_{\text{loc}}(\Omega)$ and $u$ is an $L^n$-viscosity solution of $F \leq f$. Assume that $\psi \in W^{2,n}_{\text{loc}}(\Omega)$ and $u - \psi$ has a strict maximum at $\hat{x}$; that is

$$u(x) - \psi(x) \leq u(\hat{x}) - \psi(\hat{x}) - h(r) \quad \text{for} \quad r \leq |x - \hat{x}| \leq r_0$$

where $h(r) > 0$ for $r > 0$. Since for every constant $c$

$$\begin{aligned}\mathcal{P}^-(D^2(u - \psi + c)) - \gamma|D(u - \psi + c)| &\leq F(x, u, Du, D^2u) - F(x, u, D\psi, D^2\psi) \\ &\leq f - F(x, u, D\psi, D^2\psi) \equiv g\end{aligned},$$

in the $L^n$-viscosity sense, from Proposition 2.3

$$\kappa \leq C\|g\|_{L^n(\Gamma^+_\kappa(u-\psi+c, B_r(\hat{x})))}$$

for

$$\kappa \leq \tfrac{1}{2r}\left(\sup_{B_r(\hat{x})}(u - \psi + c) - \sup_{|x-\hat{x}|=r}(u - \psi + c)\right)$$

if $c$ is chosen so that $u - \psi + c > 0$.

Using the strictness, this is satisfied for

$$\kappa \leq \tfrac{1}{2r}h(r)$$

and $r \leq r_0$. In particular, we learn that

$$\text{meas}(\Gamma^+_\kappa(u - \psi, B_r(\hat{x}))) > 0 \tag{3.1}$$

for all $\kappa$ sufficiently small given $r$ sufficiently small.

**Definition 3.1.** *$u$ has the upper perturbation property if for every $\psi \in W^{2,n}_{\text{loc}}(\Omega)$ and strict maximum $\hat{x}$ of $u - \psi$, (3.1) holds for all $0 < \kappa$ sufficiently small given $r > 0$ sufficiently small.*

Above we proved:

**Proposition 3.2.** *Let $F$ have structure, $f \in L^n_{\text{loc}}(\Omega)$ and $u$ be an $L^n$-viscosity solution of $F \leq f$ in $\Omega$. Then $u$ has the upper perturbation property.*

Similarly, supersolutions have the lower perturbation property. Note that the upper perturbation property is NOT defined by being a subsolution of some equation, although it is a property of any subsolution. Note further that any function $u \in W^{2,n}_{\rm loc}(\Omega)$ is a strong solution of an equation $-\Delta u = g \in L^n_{\rm loc}(\Omega)$ and thus has the upper perturbation property. In particular, Proposition 3.2 extends Bony's maximum principle ([12]).

We remark that the use of "pointwise a.e." together with properties like the "upper perturbation property" occurs already in Aleksandrov [1], where "absolute continuity of the normal mapping" corresponds to the upper perturbation property. Aleksandrov is aiming at global estimates, and therefore his "absolute continuity of the normal mapping" is global. Our formulation of the upper perturbation property could be strengthened in a number of ways without affecting Proposition 3.2.

**Theorem 3.3.** *Let $F$ have structure and $f \in L^n(\Omega)$. Then $u \in C(\Omega)$ is an $L^n$-viscosity solution of $F \le f$ in $\Omega$ if and only if $u$ has the upper perturbation property and is a pointwise a.e. subsolution of $F \le f$.*

*Proof.* In view of Theorem 2.5, we have only to check that the upper perturbation property and (1.6) imply that $u$ is an $L^n$-viscosity solution of $F \le f$. Let $\psi \in W^{2,n}_{\rm loc}(\Omega)$ and $\hat{x} \in \Omega$ be a strict maximum of $u - \psi$. From the upper perturbation property, for $r$ small and $z \in \Gamma^+_\kappa(u - \psi, B_r(\hat{x}))$ there exists $p \in \mathbb{R}^n$, $|p| \le \kappa$, such that $u(x) - \psi(x) - \langle p, x\rangle$ has a maximum at $z$. We may assume that $\psi$ is twice differentiable at $z$ and (1.5) holds at $x = z$ – thus

$$F(z, u(z), D\psi(z), D^2\psi(z)) - f(z) \le F(z, u(z), D\psi(z) + p, D^2\psi(z)) - f(z) + \gamma\kappa \le \gamma\kappa.$$

Since the choices of $z$ range over a set of positive measure and $\kappa$ was arbitrary, we conclude that

$$\operatorname{ess}\liminf_{z \to \hat{x}} (F(z, u(z), D\psi(z), D^2\psi(z)) - f(z)) \le 0.$$

□

We turn to the notions of Jensen. The proof we give is much more general than the statement, which we restrict to the equations treated in [10].

**Theorem 3.4.** *Let $F(X) = -\mathrm{tr}\,(A(x)X)$ have structure and $f \in L^\infty(\Omega)$. Then the following are equivalent:*

*(i) $u \in C(\Omega)$ is a $\gamma$-subsolution of $F = f$.*

*(ii) $u \in C(\Omega)$ is a $\beta$-subsolution of $F = f$.*

*(iii) $u \in C(\Omega)$ is an $L^n$-viscosity subsolution of $F = f$.*

*Proof.* Assume that $u$ is a $\gamma$-subsolution of $F = f$. If $\varphi \in C^2$ and $u - \varphi$ has a maximum at $\hat{x}$, (1.7) of Definition 1.5 implies that $\left\{\operatorname{tr}\left(A(x)D^2\varphi(x)\right) + f(x) + \eta > 0\right\} \cap B_r(\hat{x})$ has positive measure for each $\eta, r > 0$. (Note that while $\eta$ enters (1.7) in a different manner, $\lambda n \leq \operatorname{tr}(A(x)I) \leq \Lambda n$, so our claim is valid.) Then for $K \geq f$

$$-\mathcal{P}^-(D^2\varphi(x)) + K + \eta \geq \operatorname{tr}\left(A(x)D^2\varphi(x)\right) + f(x) + \eta$$

is positive on a set of positive measure in $B_r(\hat{x})$; it follows that $u$ is a $C$-viscosity solution of $\mathcal{P}^-(D^2u) \leq K$ (and hence an $L^n$-viscosity solution by Theorem 2.1). Hence $u$ has the upper perturbation property. Next, if $\hat{x}$ is a Lebesgue point of $A(x), f(x)$, we learn from (1.7) that $\operatorname{tr}\left(A(\hat{x})D^2\varphi(\hat{x})\right) + f(\hat{x}) + \eta > 0$ for $\eta > 0$ and so $-\operatorname{tr}\left(A(\hat{x})D^2\varphi(\hat{x})\right) \leq f(\hat{x})$. It now follows from Theorem 3.3 that $u$ is an $L^n$-viscosity subsolution of $F = f$ if it is a $\gamma$-subsolution.

We next show that an $L^n$-viscosity subsolution of $F = f$ is a $\beta$-subsolution. The nature of the mysterious constant $C_J$ is determined by the constants $C_{ABP}$ for which the Aleksandrov-Bakelman-Pucci maximum principle holds on balls for this equation. That is, any constant for which the inequality

$$\sup_B u \leq \sup_{\partial B} u^+ + \operatorname{diam}(B) C_{ABP} \|f^+\|_{L^n(\Gamma^+(u^+,B))}$$

is valid for $L^n$-viscosity subsolutions of the linear equation $F = f$ on any ball $B$. Suppose that $\varphi \in C^2$ and that $u - \varphi$ has a local maximum at $\hat{x} \in \Omega$. We may assume that $u(\hat{x}) - \varphi(\hat{x}) = 0$. For small $r, \eta > 0$ define

$$w(x) = u(x) - \varphi(x) - \tfrac{\eta}{2}|x - \hat{x}|^2 + \tfrac{\eta r^2}{2}.$$

Then $w$ is an $L^n$-viscosity subsolution of

$$-\operatorname{tr}\left(A(x)D^2w\right) = \operatorname{tr}\left(A(x)(D^2\varphi(x) + \eta I)\right) + f(x) \quad \text{in} \quad B_r(\hat{x})$$

and from the Maximum Principle

$$\sup_{B_r(\hat{x})} w \leq \sup_{\partial B_r(\hat{x})} w^+ + 2rC_{ABP}\left\| \left(\operatorname{tr}\left(A(x)(D^2\varphi(x) + \eta I)\right) + f(x)\right)^+ \right\|_{L^n(B_r(\hat{x}))}. \tag{3.2}$$

Since $w(\hat{x}) = \frac{\eta r^2}{2}$ and $w \leq 0$ on $\partial B_r(\hat{x})$ for small $r$, from (3.2) we obtain

$$\tfrac{\eta r^2}{2} \leq 2rC_{ABP}\left\| \left(\operatorname{tr}\left(A(x)(D^2\varphi(x) + \eta I)\right) + f(x)\right)^+ \right\|_{L^n(B_r(\hat{x}))}.$$

Next we use

$$\begin{aligned}
&\| \Big(\operatorname{tr}\big(A(x)(D^2\varphi(x)+\eta I)\big)+f(x)\Big)^+ \|_{L^n(B_r(\hat{x}))} \\
&\quad \le \left(\omega_n r^n \fint_{B_r(\hat{x})} \Big(\operatorname{tr}\big(A(x)(D^2\varphi(x)+\eta I)\big)+f(x)\Big)^+ dx\right)^{1/n} \\
&\qquad \times \| \Big(\operatorname{tr}\big(A(x)(D^2\varphi(x)+\eta I)\big)+f(x)\Big)^+ \|_{L^\infty(\Omega)}^{(n-1)/n},
\end{aligned}$$

where $\omega_n$ is the volume of the unit ball in $\mathbb{R}^n$. Gluing these things together, we may take

$$C_J = \frac{\eta^n}{4^n\omega_n \| (\operatorname{tr}(A(x)(D^2\varphi(x)+\eta I))+f(x))^+ \|_{L^\infty(\Omega)}^{(n-1)}} \frac{1}{C_{ABP}^n}.$$

□

**Remark 3.5.** It is now obvious how to generalize the definitions used by Jensen to the fully nonlinear case $F = f$, $f \in L^n(\Omega)$. One asks that

$$\tfrac{\eta r^2}{2} \le 2rC_{ABP}\| \big(f(x) - F(x,u,D\varphi+\eta(x-\hat{x}), D^2\varphi+\eta I)\big)^+ \|_{L^n(B_r(\hat{x}))},$$

where $C_{ABP}$ is now any constant depending on the structure constants $\lambda, \Lambda, \gamma$ for which the associated inequality holds. If $F$ has structure then by the same argument as above one shows that $L^n$-viscosity subsolutions of $F \le f$ in $\Omega$ are subsolutions in this new sense.

In view of Theorem 3.4, the following theorem is equivalent to the main result of [10].

**Theorem 4.1.** *Let $u \in C(\Omega)$, $\Omega$ be bounded, $F(x,X) = -\mathrm{tr}\,(A(x)X)$ have structure, and $f \in L^\infty(\Omega)$. Then $u$ is a good solution of $F = f$ in $\Omega$ if and only if $u$ is an $L^n$-viscosity solution of $F = f$ in $\Omega$.*

It follows from the definition of a good solution and Theorem 2.8 that if $u$ is a good solution, then it is an $L^n$-viscosity solution. The point is the converse, and it was surprising that this could be proved without establishing the uniqueness of good solutions of Dirichlet problems.

We review main points of Jensen's proof. Let us exhibit the issue clearly: we are concerned about an $L^n$-viscosity solution $u$ of

$$-\mathrm{tr}\left(A(x)D^2u\right) = f(x) \quad \text{in} \quad \Omega \tag{4.1}$$

in the bounded open set $\Omega$ in $\mathbb{R}^n$, where $\lambda I \le A(x) \le \Lambda I$ and $f \in L^\infty(\Omega)$. We seek to show that there are matrices $\lambda I \le A_m(x) \le \Lambda I$ and $f_m \in L^\infty(\Omega)$ and strong solutions $u_m$ of $-\mathrm{tr}\left(A_m(x)D^2u_m\right) = f_m(x)$ in $\Omega$ such that

$$A_m(x) \to A(x) \quad \text{a.e. in} \quad \Omega, \quad f_m \to f \quad \text{in} \quad L^n_{\mathrm{loc}}(\Omega), \qquad \text{and} \quad u_m \to u \quad \text{in} \quad C(\Omega). \tag{4.2}$$

The approximations $F_m(x, D^2u_m) = -\mathrm{tr}\left(A_m(x)D^2u_m\right) = f_m$ of $F = f$ are good approximations in the sense of the definition of good solutions when (4.2) holds.

The path is as follows: construct a $C^{1,1}$ approximation $\underline{u}$ of $u$ which is an $L^\infty$-strong subsolution of an approximate problem

$$-\mathrm{tr}\left(\underline{A}(x)D^2\underline{u}\right) \le \underline{f}.$$

Likewise, construct a strong supersolution $\bar{u}$ of an approximate problem

$$-\mathrm{tr}\left(\bar{A}(x)D^2\bar{u}\right) \ge \bar{f};$$

then prove that there is another problem

$$-\mathrm{tr}\left(\tilde{A}(x)D^2\tilde{u}\right) = \tilde{f}$$

admitting an exact strong solution more or less intermediate to the lower and upper approximate problems. We call this last step Jensen's In Between Theorem and do not discuss it further here – see [10], Theorem 3.30.

We present here a complete construction of the approximate strong subsolution. Our construction differs in some details from Jensen's. Our presentation is relatively self-contained, and includes improved proofs of some facts from other papers which were simply quoted in [10]. We hope that the main ideas below are more accessible here than in [10], and that our presentation is more efficient. We emphasize, however, that all the main new ideas are (sometimes variants) of Jensen's adapted to our framework.

The $C^{1,1}$ approximation process we will use is the familiar one of Lasry and Lions [11]. First we recall definitions of the sup- and inf-convolutions. Given $u\colon \mathbb{R}^n \to \mathbb{R}$, for $\epsilon > 0$ we define its sup-convolution $u^\epsilon$ and its inf-convolution $u_\epsilon$ by

$$u^\epsilon(x) = \sup_{y\in\mathbb{R}^n} \left\{u(y) - \tfrac{1}{2\epsilon}|x-y|^2\right\} \quad \text{and} \quad u_\epsilon(x) = \inf_{y\in\mathbb{R}^n} \left\{u(y) + \tfrac{1}{2\epsilon}|x-y|^2\right\}. \tag{4.3}$$

Note that the operators in (4.3) are symmetric in the sense that $(-u)^\epsilon = -u_\epsilon$. Thus for every statement about, say, sup-convolutions, there is a corresponding statement regarding inf-convolutions. In what follows we will consider various properties of convolutions and we will frequently treat only one of the two symmetric versions.

Suppose that $u\colon \mathbb{R}^n \to \mathbb{R}$ is bounded and let

$$K = \sqrt{2\text{osc}(u)}, \quad \text{osc}(u) = \sup_{\mathbb{R}^n} u - \inf_{\mathbb{R}^n} u.$$

Then for $\epsilon > 0$ and $|x-y| > K\sqrt{\epsilon}$ we have $u(y) - \frac{1}{2\epsilon}|x-y|^2 < u(x)$; consequently

$$u^\epsilon(x) = \sup\left\{u(y) - \tfrac{1}{2\epsilon}|x-y|^2\colon\ |x-y| \le K\sqrt{\epsilon}\right\}. \tag{4.4}$$

Similarly, if $u \in USC(\mathbb{R}^n)$ (the upper-semicontinuous functions on $\mathbb{R}^n$) is bounded above by a constant $C$,

$$u(x) \le u^\epsilon(x) \le \sup_{\mathbb{R}^n}\left(C - \tfrac{1}{2\epsilon}|x-y|^2\right)$$

shows that the supremum is in fact a maximum and is attained at some point $y$ such that $|x-y|^2 \le 2\epsilon(C - u(x))$.

Recall that e.g. $u^\epsilon$ is $\frac{1}{2\epsilon}$-semiconvex (meaning that $u(x)+\frac{1}{2\epsilon}|x|^2$ is convex), $u^\epsilon \to u$ uniformly whenever $u$ is uniformly continuous and that for bounded $u$, $u^\epsilon$ is globally Lipschitz with a Lipschitz constant $2\|u\|_\infty/\sqrt{\epsilon}$. Both of these assertions are easy to check directly.

Next note that for any bounded $u\colon \mathbb{R}^n \to \mathbb{R}$,

$$\inf_{\mathbb{R}^n} u \le u^\epsilon \le \sup_{\mathbb{R}^n} u \quad \text{and therefore} \quad \text{osc}(u^\epsilon) \le \text{osc}(u).$$

Given now $u \in C(\overline{\Omega})$ we extend $u$ to $\tilde{u} \in BUC(\mathbb{R}^n)$ (the uniformly continuous bounded functions on $\mathbb{R}^n$) in any way such that

$$\tilde{u}|_{\overline{\Omega}} = u, \quad \text{and} \quad \min_{\overline{\Omega}} u \leq \tilde{u} \leq \max_{\overline{\Omega}} u.$$

The approximation process we will use initially involves two parameters $0 < \epsilon, \delta$. It is

$$u \to K_{\epsilon,\delta}(u) = \left(\tilde{u}^{\epsilon+\delta}\right)_{\delta}.$$

Later we will determine $\delta$ from $\epsilon$ and other structure constants.

Observe that $(\tilde{u}^{\epsilon+\delta})_{\delta}$ is a sup-inf-convolution, as introduced by Lasry and Lions in [11]. It follows that $(\tilde{u}^{\epsilon+\delta})_{\delta} \in C^{1,1}(\mathbb{R}^n)$ and as $\epsilon, \delta \downarrow 0$, $(\tilde{u}^{\epsilon+\delta})_{\delta}$ converges uniformly to $\tilde{u}$ on $\mathbb{R}^n$, and consequently to $u$ on $\overline{\Omega}$. It will be shown below that for arbitrary bounded $w\colon \mathbb{R}^n \to \mathbb{R}$, $(w^{\epsilon+\delta})_{\delta}$ is both $\frac{1}{2\epsilon}$-semiconvex and $\frac{1}{2\delta}$-semiconcave, which is a main point in [11], and this at once implies that $(w^{\epsilon+\delta})_{\delta}$ is $C^{1,1}$. In other words, the basic results of [11] used in this paper are proved below.

**Remark 4.2.** The approximations Jensen employed in [10] are equivalent to ours in the sense that inside $\Omega$ away from the boundary, for sufficiently small $\epsilon, \delta$, $A_{\epsilon}^{+}[u] = \tilde{u}^{\epsilon}$ and $SC_{\delta}^{+}[v] = (v^{\delta})_{\delta}$, and thus $SC_{\delta}^{+}[A_{\epsilon}^{+}[u]] = (\tilde{u}^{\epsilon+\delta})_{\delta}$; see [10] for his notation. We leave the details to the interested reader.

The last assertion of Proposition 4.3 below corresponds to Lemma 2.14 in [9], but our proof is simpler. The proposition shows why sup convolutions preserve subsolution properties.

**Proposition 4.3.** *Suppose that $u \in USC(\mathbb{R}^n)$ is bounded from above, $\hat{x} \in \mathbb{R}^n$ and $\epsilon > 0$. If $(p, X) \in \mathcal{D}^{2,+}u^{\epsilon}(\hat{x})$ then for every real $n \times n$ matrix $T$*

$$\left(p, \tfrac{1}{\epsilon}(I - T^*)(I - T) + T^* X T\right) \in \mathcal{D}^{2,+}u(\hat{y}), \tag{4.5}$$

*where $\hat{y} = \hat{x} + \epsilon p$ is the unique point for which*

$$u^{\epsilon}(\hat{x}) = u(\hat{y}) - \tfrac{1}{2\epsilon}|\hat{x} - \hat{y}|^2.$$

*In particular, putting $T = I$ gives*

$$(p, X) \in \mathcal{D}^{2,+}u(\hat{y}).$$

*Moreover, if $X > \frac{1}{\epsilon}I$ then choosing $T = (I + \epsilon X)^{-1}$ in (4.5) gives*

$$\left(p, (I + \epsilon X)^{-1} X\right) \in \mathcal{D}^{2,+}u(\hat{y}).$$

*Proof.* We can find $\varphi \in C^2(\mathbb{R}^n)$ such that

$$u^\epsilon(x) - \varphi(x) \le u^\epsilon(\hat{x}) - \varphi(\hat{x}) \quad \text{for all} \quad x \in \mathbb{R}^n$$

and $(p, X) = (D\varphi(\hat{x}), D^2\varphi(\hat{x}))$. From the definition of $u^\epsilon$ there exists $\hat{y} \in \mathbb{R}^n$ such that

$$u(y) - \tfrac{1}{2\epsilon}|x-y|^2 - \varphi(x) \le u^\epsilon(\hat{x}) - \varphi(\hat{x}) = u(\hat{y}) - \tfrac{1}{2\epsilon}|\hat{x}-\hat{y}|^2 - \varphi(\hat{x}) \quad \text{for all} \quad y \in \mathbb{R}^n. \tag{4.6}$$

Taking $y = \hat{y}$ in (4.6) we find that $\frac{1}{2\epsilon}|x-\hat{y}|^2 + \varphi(x)$ has a minimum at $\hat{x}$, and therefore $\hat{y} = \hat{x} + \epsilon D\varphi(\hat{x}) = \hat{x} + \epsilon p$.

Now put $x = T(y - \hat{y}) + \hat{x}$ in (4.6). This leads to

$$u(y) - \Phi(y) \le u(\hat{y}) - \Phi(\hat{y})$$

for all $y$, where $\Phi(y) = \frac{1}{2\epsilon}|(I-T)y + T\hat{y} - \hat{x}|^2 + \varphi(Ty - T\hat{y} + \hat{x})$. We conclude that

$$\begin{aligned}\big(D\Phi(\hat{y}), D^2\Phi(\hat{y})\big) &= \big(\tfrac{1}{\epsilon}(I-T^*)(\hat{y}-\hat{x}) + T^*D\varphi(\hat{x}), \tfrac{1}{\epsilon}(I-T^*)(I-T) + T^*D^2\varphi(\hat{x})T\big) \\ &= \big(p, \tfrac{1}{\epsilon}(I-T^*)(I-T) + T^*XT\big) \in \mathcal{D}^{2,+}u(\hat{y}).\end{aligned}$$

A simple computation gives the last assertion of the proposition. □

Next two propositions correspond to Lemma 3.20 and Lemma 3.5 in [10]. To motivate the first of these, consider $w = (\tilde{u}^{\epsilon+\delta})_\delta$. We use two facts, the semigroup property

$$\tilde{u}^{\epsilon+\delta} = (\tilde{u}^\epsilon)^\delta$$

and

$$v \le (v^\delta)_\delta,$$

both of which follow quickly from the definitions. Together they imply that $\tilde{u}^\epsilon \le ((\tilde{u}^\epsilon)^\delta)_\delta = w$, which implies that $\mathcal{D}^{2,+}w(x) \subset \mathcal{D}^{2,+}\tilde{u}^\epsilon(x)$ if $w(x) = \tilde{u}^\epsilon(x)$. From Proposition 4.3 it will follow that $\tilde{u}^\epsilon$ is a subsolution of an approximation of $F = f$, and then $w$ will be as well pointwise a.e. on $\{w = \tilde{u}^\epsilon\}$. We will see that the following proposition will take care of the set $\{\tilde{u}^\epsilon < w\}$.

**Proposition 4.4.** *Let $v \in BUC(\mathbb{R}^n)$, $\delta > 0$ and suppose that $w = (v^\delta)_\delta$ is differentiable everywhere. If for some $\hat{x}$, $w(\hat{x}) > v(\hat{x})$ and $w$ is twice differentiable at $\hat{x}$, then $D^2w(\hat{x})$ has $1/\delta$ as an eigenvalue.*

*Proof.* Note that $D^2 w \le \frac{1}{\delta} I$ (in an appropriate sense) as $w$ is $\frac{1}{2\delta}$-semiconcave. This proposition shows that at $\hat{x}$ this inequality cannot be relaxed.

We have for all $x \in \mathbb{R}^n$

$$w(x) = \inf_{y \in \mathbb{R}^n} \sup_{z \in \mathbb{R}^n} \left\{ v(z) - \tfrac{1}{2\delta}|y-z|^2 + \tfrac{1}{2\delta}|x-y|^2 \right\}.$$

From Proposition 4.3 for every $x$ there is an optimal $y$, namely $y = x - \delta Dw(x)$. Hence for all $x \in \mathbb{R}^n$

$$\begin{aligned} w(x) &= \sup_{z \in \mathbb{R}^n} \left\{ v(z) - \tfrac{1}{2\delta}|z - x + \delta Dw(x)|^2 + \tfrac{1}{2\delta}|\delta Dw(x)|^2 \right\} \\ &= \sup_{z \in \mathbb{R}^n} \left\{ v(z) - \tfrac{1}{2\delta}|z-x|^2 - \langle z - x, Dw(x) \rangle \right\}. \end{aligned}$$

Since $v$ is bounded, for $x = \hat{x}$ there exists an optimal $z$, say $\hat{z}$, and then, since the map $x \mapsto w(x) + \frac{1}{2\delta}|\hat{z} - x|^2 + \langle \hat{z} - x, Dw(x) \rangle$ has a (global) minimum at $\hat{x}$,

$$\tfrac{1}{\delta}(\hat{x} - \hat{z}) + D^2 w(\hat{x})(\hat{z} - \hat{x}) = 0,$$

and consequently $\hat{z} - \hat{x}$ is an eigenvector of $D^2 w(\hat{x})$ corresponding to an eigenvalue $1/\delta$. Note that $\hat{z} \neq \hat{x}$ because otherwise we would have $w(\hat{x}) = v(\hat{x})$. Observe that $w$ being $\frac{1}{2\delta}$-semiconcave we have

$$v(\hat{z}) = w(\hat{x}) + \tfrac{1}{2\delta}|\hat{z} - \hat{x}|^2 + \langle \hat{z} - \hat{x}, Dw(\hat{x}) \rangle \ge w(\hat{z}) \ge v(\hat{z}),$$

so $w(\hat{z}) = v(\hat{z})$. □

Next we show that the operation $v \to (v^\delta)_\delta$ preserves semiconvexity, which is used twice below.

**Proposition 4.5.** *Suppose $v$ is $\frac{1}{2\kappa}$-semiconvex, $\kappa > 0$ and $\delta > 0$. Then $(v^\delta)_\delta$ is $\frac{1}{2\kappa}$-semiconvex.*

*Proof.* We begin by noting

$$\begin{aligned} (v_\kappa)^\kappa(x) &= \sup_y \inf_z \left( v(z) + \tfrac{1}{2\kappa}|z-y|^2 - \tfrac{1}{2\kappa}|y-x|^2 \right) \\ &= \sup_y \inf_z \left( v(z) + \tfrac{1}{2\kappa}|z|^2 + \tfrac{1}{\kappa}\langle y, x - z \rangle \right) - \tfrac{1}{2\kappa}|x|^2 \cdot \\ &= v(x) \end{aligned}$$

The last equality follows from the classical relation

$$g(x) = \sup_y \inf_z \{ g(z) + \langle y, x - z \rangle \}$$

whenever $g$ is convex. That the right-hand side is at most $g(x)$ follows by taking $z = x$ and that the right-hand side is at least $g(x)$ follows from

$$g(x) + \langle y^*, z - x\rangle \leq g(z)$$

whenever $y^*$ is a "subdifferential" (in the sense of convex analysis) for $g$ at $x$ – then take $y = y^*$. To proceed, we put $w = v_\kappa$ and then have $(v^\delta)_\delta = ((w^\kappa)^\delta)_\delta = (w^{\kappa+\delta})_\delta$. Finally we use the fact that if $u$ is $\frac{1}{2\alpha}$-semiconvex, $\delta < \alpha$, then $u_\delta$ is $\frac{1}{2(\alpha-\delta)}$-semiconvex (we prove this below, but see [11] p. 264). Since $w^{\kappa+\delta}$ is $\frac{1}{2(\delta+\kappa)}$-semiconvex, $(w^{\kappa+\delta})_\delta$ is $\frac{1}{2(\delta+\kappa-\delta)} = \frac{1}{2\kappa}$-semiconvex.

We prove the fact used. Suppose that $0 < \delta < \alpha$ and $u(x) + \frac{1}{2\alpha}|x|^2$ is convex. Then

$$\begin{aligned} u_\delta(x) + \tfrac{1}{2(\alpha-\delta)}|x|^2 &= \inf_y \left\{ \left(u(y) + \tfrac{1}{2\alpha}|y|^2\right) - \tfrac{1}{2\alpha}|y|^2 + \tfrac{1}{2\delta}|x-y|^2 + \tfrac{1}{2(\alpha-\delta)}|x|^2 \right\} \\ &= \inf_y \left\{ \left(u(y) + \tfrac{1}{2\alpha}|y|^2\right) + \Phi(x,y) \right\}. \end{aligned}$$

Compute the Hessian of $\Phi(x, y)$ to see that it is convex and use the simple remark that if $\Psi(x, y)$ is jointly convex in $(x, y)$, then $\inf_y \Psi(x, y)$ is convex in $x$. □

**Proposition 4.6.** *Let $u \in C(\overline{\Omega})$ be an $L^n$-viscosity subsolution of (4.1) and let $\Omega^* \subset \Omega$ be an open set whose closure lies in $\Omega$. Then there exists a constant $\mu \in (0,1)$ such that for all sufficiently small $\epsilon$ and $\delta = \mu\epsilon$*

$$-\operatorname{tr}\left(A(T_\epsilon x) D^2 w(x)\right) \leq f(T_\epsilon x) \quad a.e. \quad x \in \Omega^* \tag{4.7}$$

*holds for $w = (\tilde{u}^{\epsilon+\delta})_\delta$ and $T_\epsilon x = x + \epsilon Dw(x)$.*

*Proof.* What follows is simple, but it can be confusing, so we go slowly. First, by construction, $w$ is twice differentiable a.e. Moreover, by Theorem 2.5 and Theorem 2.6, $u$ is also twice super differentiable a.e. and is a pointwise a.e. solution of $F \leq f$. Since $\tilde{u}^\epsilon$ is $K/\sqrt{\epsilon}$-Lipschitz, so is $w$ because sup- and inf-convolutions preserve moduli of continuity. Therefore, regardless of $\delta > 0$,

$$\epsilon|Dw(x)| \leq \epsilon \frac{K}{\sqrt{\epsilon}} = K\sqrt{\epsilon}. \tag{4.8}$$

Thus there is an $\epsilon_0 > 0$ such that

$$x \in \Omega^* \Rightarrow T_\epsilon x = x + \epsilon Dw(x) \in \Omega$$

for $\epsilon < \epsilon_0$. Let $\hat{x} \in \Omega^* \backslash \mathcal{N}$ and $\epsilon < \epsilon_0$ so $T_\epsilon \hat{x} \in \Omega$. The following lemma will show that $T_\epsilon x \notin \mathcal{N}$ a.e. in $x \in \Omega^*$, so we may also assume that $T_\epsilon \hat{x} \notin \mathcal{N}$.

First suppose that $w(\hat{x}) = \tilde{u}^\epsilon(\hat{x})$. Then, since $\tilde{u}^\epsilon \le w$, $(Dw(\hat{x}), D^2w(\hat{x})) \in \mathcal{D}^{2,+}\tilde{u}^\epsilon(\hat{x})$ and by Proposition 4.3 $(Dw(\hat{x}), D^2w(\hat{x})) \in \mathcal{D}^{2,+}u(T_\epsilon\hat{x})$ and since $T_\epsilon\hat{x} \notin \mathcal{N}$ (4.7) holds for $x = \hat{x}$.

Now suppose that $w(\hat{x}) > \tilde{u}^\epsilon(\hat{x})$. From Proposition 4.4 $D^2w(\hat{x})$ has $1/\delta$ as an eigenvalue. Since $\tilde{u}^\epsilon$ is $\frac{1}{2\epsilon}$-semiconvex, by Proposition 4.5 so is $w$, and therefore all the remaining eigenvalues $\sigma$ of $D^2w(\hat{x})$ satisfy $-1/\epsilon \le \sigma \le 1/\delta$. It follows that for every $y \in \Omega$

$$-\mathrm{tr}\left(A(y)D^2w(\hat{x})\right) - f(y) \le -\lambda\tfrac{1}{\delta} + (n-1)\Lambda\tfrac{1}{\epsilon} + \|f\|_\infty, \tag{4.9}$$

which can be made negative by choosing $\delta = \epsilon\lambda/(2(n-1)\Lambda)$ and $\epsilon$ small. □

In the above proof we claimed that $T_\epsilon x \notin \mathcal{N}$ a.e. in $x \in \Omega^*$. The following lemma corresponds to Lemma 3.17 in [10] and is what we need to establish this, and more.

**Lemma 4.7.** *Under the assumptions of Proposition 4.6, there exists $\beta \in (0,1)$ independent of $\epsilon, \delta$ such that*

$$DT_\epsilon(x) \ge (1-\beta)I \quad a.e.$$

*In particular, $T_\epsilon$ is locally one-to-one, its restriction to any convex set is invertible, and the inverse of any such restriction is Lipschitz continuous with the constant $1/(1-\beta)$. In consequence, $A(T_\epsilon x)$, $f(T_\epsilon(x))$ are measurable and $T_\epsilon^{-1}$ maps null sets to null sets.*

*Proof.* Following [10], we will prove the lemma by showing that there exists $\beta \in (0,1)$, independent of $\epsilon$ and $\delta$, such that $D^2\tilde{u}^\epsilon \ge -\frac{\beta}{\epsilon}I$ pointwise a.e. and in the sense of distributions. (See sections 6.3 and 6.4 of Evans and Gariepy [7] regarding the equivalence of the pointwise and distribution senses here.) By Proposition 4.5 the same holds for $w$ and consequently $DT_\epsilon \ge (1-\beta)I$ has the asserted properties.

Suppose then that $(p, X) \in \mathcal{D}^{2,+}\tilde{u}^\epsilon(x)$, $x \in \Omega^*$. Then $X \ge -\frac{1}{\epsilon}I$, $x + \epsilon p \in \Omega$ and from Proposition 4.3 for every $\rho > 0$

$$\left(p, (I + \epsilon(X + \rho I))^{-1}(X + \rho I)\right) \in \mathcal{D}^{2,+}u(x + \epsilon p).$$

As $u$ is a $C$-viscosity subsolution of $\mathcal{P}^-(D^2u) \le C = \|f\|_\infty$ in $\Omega$, we have

$$-\frac{\Lambda n}{\epsilon} - \lambda\sum_{\sigma<0}\frac{\sigma}{1+\epsilon\sigma} \le -\Lambda\sum_{\sigma>0}\frac{\sigma}{1+\epsilon\sigma} - \lambda\sum_{\sigma<0}\frac{\sigma}{1+\epsilon\sigma} \le C, \tag{4.10}$$

where $\sigma$'s are all the eigenvalues of $X + \rho I$; note that $\sigma \ge \rho - 1/\epsilon$. Thus we have a uniform lower bound on $\epsilon\sigma/(1+\epsilon\sigma)$ for $\sigma < 0$, say $\epsilon\sigma/(1+\epsilon\sigma) > -\theta$. Then

$\epsilon\sigma \geq -\theta/(1+\theta)$ and we may take $\beta = \theta/(1+\theta)$. Taking limit as $\rho \downarrow 0$ gives the result. □

The last observations needed, before applying Jensen's In Between Theorem, are these: Since $-\mathrm{tr}\left(A(T_\epsilon x)D^2 w\right) \leq f(T_\epsilon x)$ a.e. on $\Omega^*$ in the $L^\infty$-strong sense, we may smooth up the coefficients and preserve approximation properties. Moreover, composition with $T_\epsilon$ is an approximate identity, so we do approximate our original equation. To see this, recall that $T_\epsilon$ moves points of $\Omega^*$ by at most $K\sqrt{\epsilon}$. Thus for $x, y \in \Omega^*$, if $T_\epsilon x = T_\epsilon y$ we have $|x-y| \leq 2K\sqrt{\epsilon}$. Assume that $T_\epsilon x = T_\epsilon y$ and $\epsilon$ is sufficiently small so that points within $2K\sqrt{\epsilon}$ of $\Omega^*$ lie in $\Omega$. The line segment joining $x, y$ then remains strictly inside $\Omega$ and by Lemma 4.7 we then conclude that $x = y$. We claim that

$$\lim_{\epsilon \downarrow 0} \int_{\Omega^*} |g(T_\epsilon x) - g(x)|\, dx = 0 \quad \text{for} \quad g \in L^1(\Omega);$$

this is clear for continuous $g$. For the general case, we need the boundedness of the map $g \to g \circ T_\epsilon$ in $L^1$, and this follows from

$$\int_{\Omega^*} |g(T_\epsilon x)|\, dx \leq \int_{T_\epsilon(\Omega^*)} |g(y)||\det DT_\epsilon^{-1}(y)|\, dy \leq \tfrac{1}{(1-\beta)^n} \int_{T_\epsilon(\Omega^*)} |g(y)|\, dy.$$

For the final punch line, apply [10], Theorem 3.30.

**Remark 4.8.** The approximation procedure we present here succeeds for more general equations than (4.1). Suppose that $F(x, r, p, X)$ has structure and $f \in L^\infty(\Omega)$. If $u \in C(\Omega)$ is an $L^n$-viscosity subsolution of (PDE), we claim that then $w = (\tilde{u}^{\epsilon+\delta})_\delta$ satisfies

$$F(T_\epsilon x, u(T_\epsilon x), Dw(x), D^2 w(x)) \leq f(T_\epsilon x) \quad \text{a.e.} \quad x \in \Omega^*. \tag{4.11}$$

The $w(\hat{x}) = \tilde{u}^\epsilon(\hat{x})$ case is the same as in the proof of Proposition 4.6. Instead of (4.9) we have

$$\begin{aligned} F(y, u(y), Dw(\hat{x}), D^2 w(\hat{x})) - f(y) &\leq \\ &\leq F(y, u(y), 0, 0) + \gamma|Dw(\hat{x})| + \mathcal{P}^+\left(D^2 w(\hat{x})\right) + \|f\|_\infty \\ &\leq C\left(1 + 1/\sqrt{\epsilon}\right) + \|f\|_\infty - \tfrac{\lambda}{\delta} + \tfrac{(n-1)\Lambda}{\epsilon} \leq -\tfrac{\lambda}{\delta} + \tfrac{2(n-1)\Lambda}{\epsilon} \end{aligned}$$

for small $\epsilon$, where $C$ is a constant depending on $\gamma$, $\|u\|_\infty$ and the modulus of continuity in (0.4), and again the right hand side can be made negative by an appropriate choice of the ratio $\epsilon/\delta$.

Now $u$ is a $C$-viscosity subsolution of $\mathcal{P}^{-}(D^2u) - \gamma|Du| \leq C + \|f\|_\infty$ and then (4.10) becomes

$$-\frac{\Lambda n}{\epsilon} - \lambda \sum_{\sigma<0} \frac{\sigma}{1+\epsilon\sigma} \leq C\left(1 + 1/\sqrt{\epsilon}\right)$$

and as before $\epsilon\sigma/(1+\epsilon\sigma) > -\theta$ provides an appropriate $\beta$.

To prove that (4.11) approximates (PDE) one has to impose additional assumptions on the function $x \mapsto F(x, r, p, X)$. Obtaining an analogue of Jensen's In Between Theorem is another matter.

## Section 5. References


[1] A. D. Aleksandrov, Majorization of solutions of second-order linear equations, *Amer. Math. Soc. Transl.* **68** (1968), 120–143.

[2] L. A. Caffarelli, Interior a priori estimates for solutions of fully non-linear equations, *Ann. Math.* **130** (1989), 189–213.

[3] L. Caffarelli, M. G. Crandall, M. Kocan and A. Święch, On viscosity solutions of fully nonlinear equations with measurable ingredients, to appear in *Comm. Pure Appl. Math.*

[4] M. C. Cerutti, E. B. Fabes and B. Manselli, Uniqueness of "good" solutions to the Dirichlet problem for elliptic operators with time independent coefficients, this volume.

[5] M. G. Crandall, H. Ishii and P.L. Lions, User's guide to viscosity solutions of second order partial differential equations, *Bull. Amer. Math. Soc.* **27** (1992), 1–67.

[6] L. Escauriaza, $W^{2,n}$ a priori estimates for solutions to fully non-linear equations, *Indiana Univ. Math. J.* **42** (1993), 413–423.

[7] L. C. Evans and R. F. Gariepy, **Measure Theory and Fine Properties of Functions**, CRC Press, Boca Raton, 1992.

[8] H. Ishii and P.L. Lions, Viscosity solutions of fully nonlinear second-order elliptic partial differential equations, *J. Diff. Eq.* **83** (1990), 26–78.

[9] R. Jensen, Uniqueness criteria for viscosity solutions of fully nonlinear elliptic partial differential equations, *Indiana Univ. Math. J.* **38** (1989), 629–667.

[10] R. Jensen, Uniformly elliptic pdes with bounded, measurable coefficients, to appear in *J. Fourier Anal. Appl.*

[11] J. M. Lasry and P.L. Lions, A remark on regularization in Hilbert spaces, *Israel J. Math.* **55** (1986), 257–266.

[12] P. L. Lions, A remark on Bony maximum principle, *Proc. Amer. Math. Soc.* **88** (1983), 503–508.

[13] M. V. Safonov, On a weak uniqueness for some elliptic equations, *Comm. Part. Diff. Eq.* **19** (1994), 943–957.

M.G. CRANDALL

University of California, Santa Barbara, CA 93106, USA

M. KOCAN

Australian National University, Canberra, ACT 0200, Australia

P. SORAVIA

Università di Padova, Padova, Italy

A. ŚWIĘCH

Georgia Institute of Technology, Atlanta, GA 30332, USA

D. FORTUNATO

# Morse theory and nonlinear elliptic problems

## 1. Introduction

It is well known that Morse developed his theory to study the topology of manifolds. In the last twenty years many papers appeared containing applications of Morse theory to nonlinear elliptic differential equations (see e.g. [4], [8], and their references).

The aim of this paper is to review some results in which Morse theory plays an important role in the study of some variational elliptic equations whose energy functional does not satisfy the compactness Palais-Smale condition. Moreover in the last section Morse theory is used in the study of some qualitative properties of the solutions of elliptic problems such as the number of their nodal regions.

First we shortly review some basic facts on Morse theory.

## 2. Basic facts of Morse theory

Let $V$ be an Hilbert manifold and $f$ a $C^2$ real functional on $V$. We denote by $K$ the set of the critical points of $f$

$$K = \{x \in V \mid f'(x) = 0\}.$$

Let $x \in K$ and $f''(x)$ be the second Fréchet differential of $f$ at $x$. The Morse index of $x$, denoted by $m(x, f)$ or simply by $m(x)$, is the maximal dimension (eventually $+\infty$) of a subspace $E$ of $T_xV$ (the tangent space at $x$ to $V$) on which $f''(x)$ is negative definite (i.e. for any $v \in E$, $v \neq 0$, $f''(x)[v, v] < 0$).

The large Morse index $m^*(x)$ is defined by

$$m^*(x) = m(x) + \dim Ker(f''(x)).$$

$x \in K$ is called nondegenerate if $f''(x)$ is invertible.

Let $a, b \in \mathbf{R} \cup \{+\infty\}$ and set

$$f^b = \{x \in V \mid f(x) < b\}, \; f_a^b = \{x \in V \mid a < f(x) < b\}, \; K_a^b = K \cap f_a^b.$$

If any $x \in K_a^b$ is non degenerate, the Morse polynomial of the strip $f_a^b$ is the polynomial defined by

$$M(f_a^b) = \sum_{x \in K_a^b} t^{m(x)} = \sum_q a_q t^q. \tag{2.1}$$

Observe that the coefficient $a_q$ of $t^q$ represents the number of the critical points in $f_a^b$ whose Morse index is $q$. If $K_a^b$ is infinite $M(f_a^b)$ is a formal series.

Now we define a topological invariant related to the strip $f_a^b$. The Poincaré polynomial of the strip $f_a^b$ is the polynomial defined by

$$P(f^b, f^a) = \sum_{q=0}^{\infty} \beta_q t^q \text{ where } \beta_q = \dim H_q(f^b, f^a). \tag{2.2}$$

$H_q(f^b, f^a)$ denotes the $q-th$ singular relative homology of $f^b$ with respect to $f^a$ with coefficients in some field $K$. Both in (2.1) and ( 2.2) we set $t^\infty = 0$.

The topological invariant $P(f^b, f^a)$ can be related to the Morse polynomial $M(f_a^b)$, which depends on the critical points in the strip $f_a^b$, when the functional $f$ satisfies a suitable compactness condition (namely the Palais-Smale condition). In fact the following theorem, which is the core of the Morse theory, holds:

**Theorem 2.1.** *Let $a, b \in \mathbf{R} \cup \{\infty\}$ be regular values for $f$ (i.e. if $x \in K$ then $f(x) \neq a, b$). Moreover assume the following:*

*-$f$ satisfies the Palais-Smale condition in $(a, b)$ (i.e. any sequence $\{x_n\} \subset f_a^b$ s.t. $f(x_n)$ is bounded and $f'(x_n) \to 0$ contains a convergent subsequence).*

*-Any $x \in K_a^b$ is nondegenerate and has finite Morse index.*

*Then the following equality (Morse relations) holds*

$$M(f_a^b) = P(f^b, f^a) + (1+t)Q(t) \tag{2.3}$$

*where $Q(t)$ is a formal series whose coefficients are cardinal numbers.*

Observe that the Morse relations (2.3) imply that for any $q \in \mathbf{N}$ $a_q \geq \beta_q$ (see (2.1) and (2.2)). Hence, roughly speaking, we can say that when a "topological change" of order $q$ occurs between the sublevels $f^b$ and $f^a$ (i.e. $\beta_q = \dim H_q(f^b, f^a) \neq 0$), then $f$ has at least $\beta_q$ critical points of Morse index $q$ in the strip $f_a^b$. Moreover the following result is immediately deduced from theorem 2.1.

**Corollary 2.2.** *Assume that all the hypotheses of theorem 2.1 hold. Let $y \in K_a^b$ with $m(y) < +\infty$ and assume $\beta_{m(y)} = 0$. Then $f$ has a critical point $x \in f_a^b$ with*

$$m(x) = m(y) + 1 \text{ or } m(x) = m(y) - 1. \tag{2.4}$$

Corollary 2.2 is immediately deduced from the Morse relations exploiting the form of the "remainder term" $(1+t)Q(t)$. Roughly speaking corollary 2.2 says that when there is a critical point $y$ which "does not come from the change of topology" (i.e. $\beta_{m(y)} = 0$), then another critical point $x$ appears whose Morse index $m(x)$ satisfies (2.4). Now we describe, by means of a simple example, this principle of coupling of critical points which are not related to "changes of topology".

Consider the 2-dimensional sphere $S^2 = \{(x, y, z) \in \mathbf{R}^3 \mid x^2 + y^2 + z^2 = 1\}$ and the height functional $f$ on $S^2$ (i.e. $f(x, y, z) = z$).

Then it is well known that the Poincaré polynomial of $S^2$ is

$$P(S^2) = 1 + t^2. \tag{2.5}$$

Moreover, since $f$ has only two critical points on $S^2$, namely the north pole $N$ and the south pole $S$, the Morse polynomial of $f$ on $S^2$ clearly is

$$M(f, S^2) = t^{m(S,f)} + t^{m(N,f)} = 1 + t^2. \tag{2.6}$$

Comparing (2.5) with (2.6), we deduce that the Morse relations (2.3) are satisfied if we take $Q(t) = 0$.

Now let us "continuously deform" $S^2$ into a new surface $\tilde{S}$ in such a way to get another critical point $A$ which is a local minimum point of $f$ on $\tilde{S}$ (see the figure). Clearly $\tilde{S}$ has the same topology of $S^2$, then

$$P(\tilde{S}) = P(S^2) = 1 + t^2. \tag{2.7}$$

Then, since the Morse polynomial of $f$ on $\tilde{S}$ "contains" (besides $1 + t^2$) the monomial $t^{m(A,f)} = 1$, by (2.7) and using the Morse relations for $f$ on $\tilde{S}$, we immediately deduce that $f$ has another critical point $B$ on $\tilde{S}$ with Morse index $m(B, f) = 1$ (see the figure).

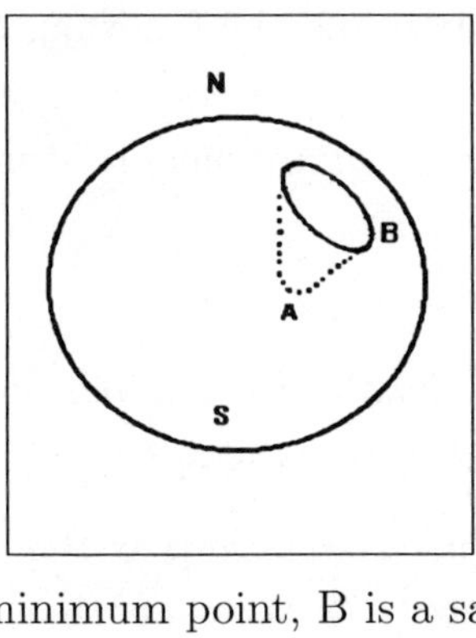


A is local minimum point, B is a saddle point.

## 3. Asymptotically linear elliptic equations at resonance

Consider the problem

$$-\Delta u - au = g(u) \text{ in } \Omega, \;\; u = 0 \text{ on } \partial\Omega \tag{3.1}$$

$\Omega$ is a bounded domain in $\mathbf{R}^n$, $a \in \mathbf{R}$ and $g$ is a bounded, $C^1$ real function such that $g(0) = 0$.

We assume that problem (3.1) is asymptotically linear i.e.

$$\lim_{|t| \longrightarrow +\infty} g'(t) = 0. \tag{3.2}$$

So we can consider the linearized operators $L_\infty$ and $L_0$ at infinity and zero defined by

$$\text{fo } u \in H_0^1(\Omega\,)\; L_\infty u = -\Delta u - au, \;\; L_0 u = -\Delta u - au - g'(0)u.$$

Clearly the solutions of (3.1) are the critical points of the energy functional

$$f(u) = \frac{1}{2} < L_\infty u, u > - \int_\Omega G(u(x))dx, \;\; u \in \; H_0^1(\Omega), \tag{3.3}$$

where $<,>$ denotes the pairing between the Sobolev space $H_0^1(\Omega)$ and its dual $H^{-1}$ and

$$G(t) = \int_0^t g(s)ds. \tag{3.4}$$

Assume first that no resonance at infinity occurs, i.e. we assume that the real number $a$ is not an eigenvalue of $-\Delta$ (with zero Dirichlet boundary data).

In this case $\ker L_\infty = \{0\,\}$ and it is not difficult to show that all the solutions of (3.1) are uniformely bounded in the $H_0^1(\Omega)$ norm. This boundness implies that (3.3) satisfies the Palais-Smale condition and the following theorem can be proved ([1])

**Theorem 3.1.** *Under the above assumptions problem (3.1) has a nontrivial solution if*

$$m(0) \neq m(\infty). \tag{3.5}$$

*Here $m(0)$ (respectively $m(\infty)$) denotes the number of the negative eigenvalues (counted according to their multiplicity) of $L_0$ (respectively $L_\infty$).*

Assumption (3.5) amounts to require that the nonlinear term $g$ interacts with the spectrum of $L_\infty$, more precisely (3.5) means that there is at least one eigenvalue of $L_\infty$ between $g'(0)$ and $g'(\infty)(= 0)$.

Assume now that resonance at infinity occurs, i.e.

$$Ker L_\infty \neq \{0\} \tag{3.6}$$

and that

$$G(t) = \int_0^t g(s)ds \text{ is bounded.} \tag{3.7}$$

Assumptions (3.6), (3.7) imply that the energy functional (3.3) does not satisfy the Palais-Smale condition. In order to see this we consider a sequence $\{u_n\} \subset Ker L_\infty$ such that

$$\lim_{n \longrightarrow +\infty} \| u_n \| = \infty \tag{3.8}$$

and we prove that it is a Palais-Smale sequence, i.e. $f(u_n)$ is bounded and $f'(u_n) \to 0$ (such a sequence is called "critical point at infinity" for our functional).

Clearly $f(u_n)$ is bounded since $G$ is bounded. Now (3.2) and (3.7) imply that

$$\lim_{|t| \longrightarrow +\infty} g(t) = 0. \tag{3.9}$$

By lemma 3.2 in ([3]), (3.8) and (3.9) imply that

$$\int_\Omega g(u_n)^2 dx \to 0 \text{ as } n \to \infty. \tag{3.10}$$

On the other hand we have for all $v \in H_0^1(\Omega)$

$$|< f'(u_n), v >|^2 \leq (\int_\Omega g(u_n)^2 dx)(\int_\Omega v^2 dx). \tag{3.11}$$

Clearly (3.10) and (3.11) imply that $f'(u_n) \to 0$.

Several existence results of nontrivial solutions of problem (3.1), under the assumptions (3.6) and (3.7), have been obtained under the additional assumption of "strong resonance" at infinity, i.e. assuming that $G(t)$ has a limit $c \in \mathbf{R}$ as $| t |$ goes to infinity. It can be shown that under this assumption the Palais-Smale condition fails to hold only at level $c$ (i.e. only for those sequences $\{u_n\}$ such that $f(u_n) \to c$). Then, by using minimax methods, it is possible to avoid the "bad" level $c$ if $G$ satisfies suitable sign conditions (see e. g.[3] and [11] for some of the first results in the strong resonance case).

Here we want to study problem (3.1) without the strong resonance assumption. Then there is not a precise level at which the Palais-Smale condition fails to hold.

A resonance case of this nature has been recently analyzed in [6] in the study of a class of asymptotically linear dynamical systems with bounded potentials.

The method developed in [6] can be usefully used to study the nonlinear elliptic problem (3.1). In fact in a recent paper A.Masiello and L.Pisani proved ([9]) the following result

**Theorem 3.2.** *Assume that (3.2), (3.6), and (3.7) hold and that* $a + g'(0)$ *is not eigenvalue of* $-\Delta$ *i.e.*

$$KerL_0 = \{0\}. \tag{3.12}$$

*Assume moreover that*

$$m(\infty) - m(0) > \text{ or } m(0) - m^*(\infty) > 1, \tag{3.13}$$

*where* $m(\infty)$, $m(0)$ *have been introduced in theorem 3.1, and*

$$m^*(\infty) = m(\infty) + \dim KerL_\infty.$$

*Under the above assumptions (3.1) has a nontrivial solution.*

Observe that assumption (3.13) means that there are at least two eigenvalues (counted with their multiplicity) of $L_\infty$ between $g'(0)$ and $g'(\infty) = 0$. Then, comparing (3.13) with (3.5), we can say that when resonance at infinity occurs a "stronger interaction" between the spectrum of $L_\infty$ and the nonlinear term $g$ is required in order to overcome the difficulties arising from the lack of the Palais-Smale condition.

## 4. Sketch of the proof of the theorem

We confine ourself to examine the case in which

$$m(\infty) - m(0) > 1. \tag{4.1}$$

We recall that the functional $f$ does not satisfies the Palais-Smale condition since any sequence $\{u_n\} \subset KerL_\infty$, such that $\| u_n \| \to \infty$, is a critical point at infinity. To avoid this difficulty we add a penalizing term which is "sensitive" on those $u$ whose component on $KerL_\infty$ is "large". More precisely we set

$$\Psi_R(u) = (\| Pu \|^2 - R)^4 \text{ if } \| Pu \|^2 \geq \text{ and } \Psi_R(u) = 0 \text{ otherwise},$$

where $R > 0$ and $P$ denotes the projection on $KerL_\infty$.

We shall consider the penalized functional

$$f_R(u) = f(u) - \Psi_R(u), \; u \in H_0^1(\Omega). \tag{4.2}$$

It is easy to show that $f_R$ satisfies the Palais-Smale condition.

Now we can divide the proof of the theorem in two steps:

**Step 1.**

First we provide a control from below on the Morse indices of the critical points of $f_R$ whose components in $Ker L_\infty$ are "large". More precisely we prove the following lemma

**Lemma 4.1.** *There exists $Q > 0$ (independent on $R$) such that, if $u$ is a critical point of $f_R$ with $\| Pu \|^2 \geq Q$, then*

$$m(u) > m(0) + 1. \tag{4.3}$$

Now we give an idea of the proof of this lemma.

Let $u$ be a critical point of $f_R$ and let $H^-$ denote the negative span of $L_\infty$. Then, since $\Psi_R$ is convex, it is easy to check that for all $v \in H^-$ we have

$$f_R''(u)[v,v] \leq f''(u)[v,v] \leq -c_1 \| v \|^2 + c_2 \| v \|^2 \| g'(u) \|_{L^{\frac{n}{2}}} \tag{4.4}$$

where $c_1$, $c_2$ are positive constants.

Since $g'(t) \to 0$ for $| t | \to \infty$, by using lemma 3.2 in [3], we can deduce that

$$\| g'(u) \|_{L^{\frac{n}{2}}} \to 0 \ for \ \| Pu \|^2 \to \infty \ . \tag{4.5}$$

Then from (4.4) and (4.5) we deduce that

$$m(u) \geq \dim H^- = m(\infty) \text{ if } \| Pu \|^2 \text{ is } l \arg \text{ e}.$$

The above inequality and assumption (4.1) imply (4.3).

**Step 2.**

Take $R = Q$ (see lemma 4.1) and consider the penalized functional $f_Q$4.1, a nontrivial critical point $u$ of $f_Q$ with $m(u) \leq m(0) + 1$ is also a critical point of $f$.

To make more clear the proof of the existence of such critical points we assume that all the critical points of $f_Q$ are nondegenerate so we can use corollary 2.2 (the nondegeneracy assumption can be avoided if we use a generalized Morse theory [7]).

Roughly speaking the functional $f_Q$ "goes to $-\infty$ along $m^*(\infty)$ directions". This fact permits to evaluate, if $a < 0$ and $| a |$ is sufficiently large, the Poincaré polynomial

$$P(f_Q^\infty, f_Q^a) = P(H_0^1(\Omega), f_Q^a) = t^{m^*(\infty)}. \tag{4.6}$$

Since 0 is a critical point of $f_Q$ with $f_Q(0) = 0 > a$ and $m(0) \neq m^*(\infty)$ (see (4.1)), we immediately deduce from (4.6) and corollary 2.2 that $f_Q$ has a critical point $u \neq 0$ with $m(u) \leq m(0) + 1$.

## 5. Nodal regions

Let $\Omega$ be an open, bounded and connected set in $\mathbf{R}^n$. For $k \in \mathbf{N}$ consider the problem

$$- \Delta u - \lambda u - \mid u \mid^{p-2} u = 0 \text{ in } \Omega \ , \ u = 0 \text{ on } \partial\Omega \tag{5.1}$$

where $2 < p < \frac{2n}{n-2}$ if $n \geq 3$ and $2 < p$ if $n = 1$ or 2. Moreover we take

$$\lambda_k \leq \lambda < \lambda_{k+1}. \tag{5.2}$$

Here $\lambda_1 < \lambda_2 \leq ... \leq \lambda_i \leq ...$ denote the eigenvalues of $-\Delta$ (with zero Dirichlet boundary conditions) repeated according to their multiplicity.

When $k = 0$ (5.2) means that $\lambda < \lambda_1$. Let $u$ be a solution of (5.1). Any open, connected subset $A \subset \mathbf{R}^n$ such that $u$ does not change sign in $A$ and $u = 0$ on $\partial A$ is called nodal region of $u$.

When $k = 0$, i.e. when $\lambda < \lambda_1$, (5.1) has a positive solution that is a solution with exactly one nodal region (see[10]). When $k \geq 1$ (5.1) does not possess positive solutions, nevertheless it possible to obtain an upper bound to the number of the nodal regions via Morse theory. More precisely the following theorem holds (see [5])

**Theorem 5.1.** *Problem (5.1) possess one nontrivial solution $u$ with*

$$Z(u) \leq k + 1, \tag{5.3}$$

*$Z(u)$ being the number of the nodal regions of $u$.*

Now we give a sketch of the proof of theorem 5.1.

The solutions of (5.1) are the critical points of the energy functional

$$f(u) = \frac{1}{2} \int_\Omega (\mid \nabla u \mid^2 - \lambda \mid u \mid^2) dx - \frac{1}{p} \int_\Omega \mid u \mid^p dx, \ u \in H_0^1(\Omega).$$

Clearly 0 is a critical point of $f$ and its Morse index $m(0)$ is given by the number of the negative eigenvalues of the operator $L_0 v = -\Delta u - \lambda u$.

Then by (5.2) we have

$$m(0) \leq k. \tag{5.4}$$

To make more clear the proof we assume that all the critical points are nondegenerate, then we can apply directly corollary 2.2.

First we observe that, by the assumptions on $p$ , standard computations permit to show that the functional $f$ satisfies the Palais-Smale condition.

Since $p > 2$, we can deduce that, roughly speaking, the functional $f$ "goes to $-\infty$ in infinitely many directions". This fact permits to show that (see [4]), if $a < 0$, we have

$$P(f^\infty, f^a) = P(H_0^1(\Omega), f^a) = t^\infty = 0.$$

So the monomial $t^{m(0)}$ is not "contained" in $P(f^\infty, f^a)$ and corollary 2.2 implies that there exists a critical point $u$ with Morse index

$$m(u) = m(0) + 1 \text{ or } m(u) = m(0) - 1. \tag{5.5}$$

Then $u \neq 0$ and by (5.4) and (5.5) we have

$$m(u) \leq k + 1. \tag{5.6}$$

On the other hand (see also [2])

$$Z(u) \leq m(u). \tag{5.7}$$

In fact let $A_i$ be a nodal region of $u$ and set

$$v_i = u \text{ in } A_i \text{ and } v_i = 0 \text{ in } \Omega \backslash A_i.$$

Then a direct calculation shows that

$$< f''(u)v_i, v_i >< 0.$$

So any nodal region of $u$ gives rise to a "negative direction" for $f''(u)$, then (5.3) follows from (5.6) and (5.7).

**References**


[1] Amann, H., Zehnder, E., Nontrivial solutions for a class of nonresonance Problems and Applications to Nonlinear Differential equations, *Ann. Sc. Norm. Sup. Pisa* **VII** (1980), 539-603.

[2] Bahri, A., Lions, P.L., Solutions of Superlinear Elliptic Equations and their Morse Indices, *Comm. Pure Appl. Math.* **XLV** (1992), 1205-1215.

[3] Bartolo, P., Benci, V., Fortunato, D., Abstract critical point theorems and applications to some nonlinear problems with strong resonance at infinity, *J. Nonlin. Anal., T. M. A.*, **7** (1983), 981-1012.

[4] Benci, V., A new approach to the Morse-Conley theory and some applications, *Ann. Mat. Pura Appl.*, **158** (1991), 231-305.

[5] Benci, B., Fortunato, D., A remark on the nodal regions of some superlinear elliptic equations, *Proc. Royal Soc. Edinburgh*, **111 A** (1988), 123-128.

[6] Benci, V., Fortunato, D., Periodic solutions of asymptotically linear dynamical systems, *Nonlinear Diff. Eq. Appl.*, **2** (1994),267-280.

[7] Benci, V., Giannoni, F., Morse theory for $C^1$ functionals and Conley blocks, to appear.

[8] Chang, K.C., Infinite dimensional Morse theory and multiple solution problem, Birkhäuser, Boston-Basel, 1993.

[9] Masiello, A., Pisani, L., Asymptotically Linear Elliptic Problems at Resonance, *Ann. Mat. Pura Appl.*, to appear.

[10] Lions, P.L., On the existence of positive solutions of semilinear elliptic equations, *SIAM Rev.* **24** (1982), 441-467.

[11] Thews, K., Nontrivial solutions of elliptic equations at resonance, *Proc. R. Soc. Edinb.* **85 A** (1980), 119-129.

Donato Fortunato

Dipartimento di Matematica

Università di Bari - 70125 Bari, Italy

E. LANCONELLI

# On a class of Kolmogorov-Fokker-Planck operators

We report on some new results recently obtained by M. Manfredini, S. Polidoro and E. Lanconelli on a class of ultraparabolic operators of Kolmogorov-Fokker-Planck type. In particular we present local and global estimates of the fundamental solution, Hölder interior estimates, and finally some existence results for the first boundary value problem.

**1. Introduction.** The aim of this paper is to present some recent results on operators of the type

$$\begin{aligned} L &= \operatorname{div}(A(z)D) + \langle x, BD\rangle - \partial_t \\ &\equiv \operatorname{div}(A(z)D) + Y \end{aligned} \tag{1.1}$$

where $z = (x,t) \in \mathbf{R}^{N+1} \equiv \mathbf{R}^N_x \times \mathbf{R}_t$, div is the divergence operator in $\mathbf{R}^N$, $D \equiv (\partial_{x_1}, \ldots, \partial_{x_N})$, is the gradient operator in $\mathbf{R}^N$ and $\langle\cdot,\cdot\rangle$ is the inner product in $\mathbf{R}^N$; $B = (b_{i,j})$ is a constant $N \times N$ matrix and $A(z) = (a_{i,j}(z))_{i,j=1,\ldots,N}$ is symmetric and non negative defined, with constant rank $p_0$ satisfying

$$A(z) = \begin{pmatrix} A_0(z) & 0 \\ 0 & 0 \end{pmatrix}, \qquad A_0(z) = (a_{i,j}(z))_{i,j=1,\ldots,p_0}.$$

Moreover, $A_0(z)$ is uniformly positive defined on $\mathbf{R}^{p_0}$, i.e. there exists a positive constant $\mu > 0$ such that

$$\frac{1}{\mu}\langle\xi,\xi\rangle \le \langle\xi, A_0(z)\xi\rangle \le \mu\langle\xi,\xi\rangle, \quad \xi \in \mathbf{R}^{p_0}, \quad 1 \le p_0 \le N.$$

Operators of type 1.1 arise, for instance, in the theory of diffusion processes from a probabilistic point of view. For example, when $N = 2n$,

$$A(z) = \begin{pmatrix} I_n & 0 \\ 0 & 0 \end{pmatrix}, \quad \text{and} \quad B = \begin{pmatrix} 0 & I_n \\ 0 & 0 \end{pmatrix},$$

the operator 1.1 becomes

$$L = \sum_{j=1}^{n} \partial_{x_i}^2 + \sum_{j=1}^{n} x_j \partial_{x_{n+j}} - \partial_t \tag{1.2}$$

which is the prototype of the Kolmogorov operator; see [S]. In this case, the solution $u$ of the equation $Lu = 0$, describes the probability density of a system with $N$ degrees of freedom.

It can be also noticed that, again for $N = 2n$, the operator 1.1 is a linearized prototype of the Fokker-Planck equation

$$\sum_{j=1}^{n} \partial_{x_j}(a(u)\partial_{x_j}u + b(u)u) + \sum_{j=1}^{n} x_j \partial_{x_{n+j}} u - \partial_t u = 0$$

which arises in the study of the Brownian motion of a particle in a fluid [C].

The operator $L$ in 1.1 is elliptic-parabolic. However, the classical techniques and results of Picone, Fichera, Khon-Nirenberg and Olěinik-Radkěvic seem not to be suitable for a satisfactory study of the properties of $L$.

We thus propose a different approach, which is based on the following idea. We first study the *model operator*

$$L = \operatorname{div}(AD) + Y, \tag{1.3}$$

where $A$ is an $N \times N$ constant matrix. Then, we analyze the variable coefficients case $A = A(z)$ as a perturbation of 1.3. In other words, our basic strategy consists of considering $L$ in 1.3 as a model operator for $L$ in 1.1, the way $\Delta - \partial_t$ is a model for the classical parabolic operator.

**2. Model operator.** In this section we consider the model operator 1.3 with constant matrix $A$. In order to find a fundamental solution of $L$ we may try to solve the Cauchy problem

$$\begin{cases} Lu = 0 & \text{for } t > 0 \\ u\,|_{t=0} = \phi. \end{cases}$$

By means of the Fourier Transform in the "spatial variable" $x$, we find

$$\hat{u}(\xi, t) = \hat{\phi}(E(t)) \exp\left(\langle C(t)\xi, \xi\rangle - t\,\mathrm{tr}(B)\right) \tag{2.1}$$

where

$$C(t) = \int_0^t E(s)AE(s)^T ds, \quad E(s) = \exp(-sB^T).$$

In general the weak condition $C(t) \geq 0$ holds, which implies that 2.1 cannot be inverted. Obviously, in the parabolic case where $A > 0$, we have $C(t) > 0$ for all $t > 0$. Nonetheless, it is important to observe that in some other special cases we may have $C(t) > 0$ for all $t > 0$ even for singular matrix $A$. For example, for the Kolmogorov operator in $\mathbf{R}^2 \times \mathbf{R}$

$$L = \partial_{x_1}^2 + x_1 \partial_{x_2} - \partial_t,$$

it holds

$$C(t) = \begin{pmatrix} t & \frac{t^2}{2} \\ \\ \frac{t^2}{2} & \frac{t^3}{3} \end{pmatrix},$$

which yields $C(t) > 0$ for all $t > 0$, since

$$\det(C(t)) = \frac{t^4}{12} > 0, \quad \forall t > 0.$$

We next assume that

$$C(t) > 0, \qquad \forall t > 0. \tag{2.2}$$

Under this hypothesis, the Fourier transform in 2.1 can be inverted, obtaining

$$u(x,t) = \int_{\mathbf{R}^N} \phi(\xi) \Gamma(x - E(t)\xi, t) d\xi$$

with $\Gamma(x,t) = 0$, for $t \leq 0$ and

$$\Gamma(x,t) = \frac{(4\pi)^{-\frac{N}{2}}}{\sqrt{\det C(t)}} \quad \exp(-\frac{1}{4}\langle C^{-1}(t)x, x\rangle - t \ \mathrm{tr}(B))$$

for $t > 0$. The function

$$\Gamma(z, \zeta) = \Gamma(x, t, \xi, \tau) = \Gamma(x - E(t-\tau)\xi, t - \tau)$$

is a fundamental solution for the operator $L$ [K]. We explicitly observe that

$$\Gamma \in C^\infty(\{(z,\zeta) \,/\, z \neq \zeta\})$$

from which it readily follows that $L$ is *hypoelliptic*. The requirement $C(t) > 0$ is not only sufficient for hypoellipticity, but it is also necessary, as it is shown in the following proposition.

**Proposition 2.1** *The following statements are equivalent.*

(*i*) *$L$ is hypoelliptic;*

(*ii*) *$C(t) > 0, \forall t > 0$;*

(*iii*) *(Hörmander condition [H])*

$$\operatorname{rank} \mathcal{L}(X_1, \ldots, X_N, Y) = N + 1,$$

*where $X_j = \sum_{i=1}^N a_{i,j} \partial_{x_i}$, $Y = \langle x, BD \rangle - \partial_t$, in which $\mathcal{L}(X_1, \ldots, X_N, Y)$ denotes the Lie algebra generated by $X_1, \ldots, X_N, Y$.*

(*iv*) *There exists a basis in $\mathbf{R}^N$ for which the matrices $A$ and $B$ take the form ([LP])*

$$A = \begin{pmatrix} A_0 & 0 \\ 0 & 0 \end{pmatrix}, \quad B = \begin{pmatrix} * & B_1 & 0 & \cdots & 0 \\ * & * & B_2 & \cdots & 0 \\ \vdots & \ddots & \ddots & \ddots & \ddots \\ * & * & * & * & B_r \\ * & * & * & * & * \end{pmatrix},$$

*where $A_0 > 0$ in $\mathbf{R}^{p_0}$, and, for $i = 1, \ldots, r$, $B_i$ is a $p_{i-1} \times p_i$ matrix with $\operatorname{rank}(B_i) = p_i$, $p_{i-1} \geq p_i$ and $p_0 + p_1 + \ldots + p_r = N$.*

We next give some examples of operators that satisfy the hypotheses above.

*Example 1.* The prototype of the Kolmogorov operator in 1.2 is in its canonical form with respect to the canonical basis in $\mathbf{R}^N$.

*Example 2.* The operator $L = \partial_{x_1}^2 + x_1 \partial_{x_2} + \ldots + x_{N-1} \partial_{x_N} - \partial_t$ in $\mathbf{R}^N \times \mathbf{R}$ is hypoelliptic. In fact, $L$ is written in its canonical form, again with respect to the canonical basis of $\mathbf{R}^N$. The corresponding matrices $A$ and $B$ are, respectively, given by

$$A = \begin{pmatrix} 1 & 0 & \ddots \\ 0 & 0 & \ddots \\ \vdots & \ddots & \ddots \\ 0 & 0 & 0 \end{pmatrix}, \quad B = \begin{pmatrix} 0 & 1 & 0 & \cdots & 0 \\ 0 & 0 & 1 & \cdots & 0 \\ \vdots & \ddots & \ddots & \ddots & \ddots \\ 0 & 0 & 0 & 0 & 1 \\ 0 & 0 & 0 & 0 & 0 \end{pmatrix}.$$

It is remarkable that a natural geometric-algebraic structure can be defined for the operator $L$ in $\mathbf{R}^N \times \mathbf{R}$. This allows us to treat $L$ as if it had all constant coefficients, including those of the first order derivatives. The following results were proved in [[LP]].

*Proposition 2.2.*

1. *$L$ is invariant with respect to the left translations of the (Lie) group $(\mathbf{R}^{N+1}, \circ)$, where the composition law is defined as*

$$z \circ z' \equiv (x,t) \circ (x',t') = (x' + E(t'+t)x, t'+t)$$

2. *Let $D(\lambda)$ be the dilation in $\mathbf{R}^N \times \mathbf{R}$ defined as*

$$D(\lambda) : \mathbf{R}^{N+1} \to \mathbf{R}^{N+1}, \qquad D(\lambda)(x,t) = (\lambda^{\alpha_1} x_1, \dots, \lambda^{\alpha_N} x_N, \lambda^{\beta} t).$$

*Then there are $\alpha_1, \dots, \alpha_N, \beta \in \mathbf{R}^+$ such that*

$$L \circ D(\lambda) = \lambda^2 D(\lambda) \circ L$$

*if and only if all '*' blocks of $B$ are zero matrices. In this case*

$$D(\lambda) = \mathrm{diag}(\lambda I_{p_0}, \lambda^3 I_{p_1}, \dots, \lambda^{2r+1} I_{p_r}, \lambda^2)$$

In other words, property 2. of Proposition 2.2 says that $L$ commutes with the dilation $(D(\lambda))_{\lambda>0}$.

*Example:* The dilation group for which the Kolmogorov operator in $\mathbf{R}^3$ is invariant is given by

$$D(\lambda) = (\lambda, \lambda^3, \lambda^2).$$

The previous result suggests the following definition, which introduces the notion of principal part of an operator $L$.

**Definition 2.3.** *Given an operator $L = \mathrm{div}(AD) - \langle x, BD \rangle - \partial_t$, we call principal part of $L$ the operator*

$$L_0 = \mathrm{div}(AD) - \langle x, B_0 D \rangle - \partial_t$$

*where $B_0$ is obtained by zeroing all '*' blocks of $B$.*

This seems to be a "natural" definition, in light of the following proposition.

*Proposition 2.4. Let $\Gamma$ denote the fundamental solution of $L$ and $\Gamma_0$ denote the fundamental solution of $L_0$. Then, for all $b > 0$ there exists $c = c(b) > 0$ such that*

$$c\,\Gamma_0(y) \le \Gamma(y) \le \frac{1}{c}\,\Gamma_0(y), \quad y = (x,t)$$

*on the level set $\{\Gamma_0(y) > b\}$. Moreover, $c(b) \to 1$, as $b \to \infty$.*

**3. Operators with variable coefficients.** Given the matrix $A_0(z) = (a_{i,j})_{i,j=1,\dots,p_0}$ $(1 \le p_0 \le N)$, in this section we consider operators of type

$$L = \sum_{i,j=1}^{p_0} a_{i,j}(z)\partial_{x_i x_j} + \langle x, BD\rangle - \partial_t \qquad \text{in} \quad \mathbf{R}^{N+1}$$

where, as in Section 1,

$$\frac{1}{\mu}\langle \xi, \xi\rangle \le \langle \xi, A_0(z)\xi\rangle \le \mu\langle \xi, \xi\rangle, \quad \forall \quad \xi \in \mathbf{R}^{p_0} \tag{3.1}$$

for all $z \in \mathbf{R}^N \times \mathbf{R}$. In addition, $B$ is supposed to be a constant $N \times N$ matrix having pattern

$$B = \begin{pmatrix} 0 & B_1 & 0 & \cdots & 0 \\ 0 & 0 & B_2 & \cdots & 0 \\ \vdots & \ddots & \ddots & \ddots & \ddots \\ 0 & 0 & 0 & 0 & B_r \\ 0 & 0 & 0 & 0 & 0 \end{pmatrix}, \tag{3.2}$$

where, for $i = 1, \dots, r$, $B_i$ is a $p_{i-1} \times p_i$ matrix of rank $p_i$, with $p_{i-1} \ge p_i$, and $p_0 + \ldots + p_r = N$.

The results of the previous section guarantee that, under these hypotheses, all *frozen operators*

$$L_\zeta = \sum_{i,j=1}^{p_0} a_{i,j}(\zeta)\partial_{x_i x_j} + \langle x, BD\rangle - \partial_t$$

are hypoelliptic. It can also be shown that $L_\zeta$ is invariant with respect to left translation of the group

$$(\mathbf{R}^N, \circ), \quad (x,t) \circ (x', t') = (x' + E(t+t')x, t+t')$$

and it commutes with the dilation

$$D(\lambda) = \operatorname{diag}(\lambda I_{p_0}, \dots, \lambda^{2r+1} I_{p_r}, \lambda^2).$$

It is important to remark that the groups $(\mathbf{R}^N, \circ)$, and $(D(\lambda))_{\lambda>0}$ do not depend on $\zeta$, but only on the matrix $B$.

Next theorem states that, assuming Hölder continuity of the coefficients $a_{i,j}$ with respect to the "geometry" just recalled, there exists a fundamental solution for $L$ which can be estimated in terms of the fundamental solution of $L_\zeta$.

**Theorem 3.1**[[P1]] *Suppose that conditions 3.1 and 3.2 are satisfied. Moreover, assume that there exist two positive constants $M$ and $\alpha$, $0 < \alpha \le 1$ such that*

$$|a_{i,j}(z) - a_{i,j}(\zeta)| \ \le \ M\|\zeta^{-1} \circ z\|_B^\alpha, \qquad \text{where} \qquad \|(x,t)\|_B = \sum_{j=1}^N |x_j|^{\frac{1}{\alpha_j}} + |t|^{\frac{1}{2}}$$

*in which the $\alpha_j$ are as defined in Proposition 2.2. Then*

(*a*) *There exists a fundamental solution $\Gamma(z,\zeta)$ of $L$ in $\mathbf{R}^{N+1} \times \mathbf{R}^{N+1}$ and it can be determined using the parametrix method of Levi;*

(*b*) *Fixed $\zeta \in \mathbf{R}^{N+1}$ and letting $Z(z;\zeta)$ be the fundamental solution of $L_\zeta$ with pole at $\zeta$, then, for all $b > 0$ there exists a positive constant $c$ such that*

$$\frac{1}{c}Z(z;\zeta) \le \Gamma(z,\zeta) \le cZ(z,\zeta)$$

*on the level set $\{z : \ Z(z;\zeta) > b\}$*

We should recall that the result (*a*) in Proposition was proved under additional regularity hypotheses on the coefficients by M. Weber [W], A.M. Il'in [I] and I. M. Sonin [So].

For operators in divergence form, in addition to the existence of the fundamental solution $\Gamma$, we next show that it is possible to derive global lower and upper estimates for $\Gamma$. This result was proved by Polidoro ([P2]) using a technique reminiscent of the *Aronson-Serrin* method related to the classical parabolic operator [AS].

**Theorem 3.2**[[P2]] *If $L = \sum_{i,j=1}^{p_0} \partial_{x_i}(a_{i,j}(z)\partial_{x_j}) + \langle x, BD\rangle - \partial_t$ where, for $i,j,k = 1,\ldots,p_0$, $a_{i,j}$, $\partial_{x_k}a_{i,j}$ are B-hölder continuous in $\mathbf{R}^{N+1}$, then there exists a fundamental solution of $L$ in $\mathbf{R}^{N+1} \times \mathbf{R}^{N+1}$. Moreover, for every fixed $T_1, T_2 \in \mathbf{R}$, with $T_1 < T_2$, there exists a positive constant $c$ such that*

$$\frac{1}{c}\Gamma^-(z;\zeta) \le \Gamma(z,\zeta) \le c\Gamma^+(z;\zeta) \qquad \text{in} \quad \mathbf{R}^N \times ]T_1, T_2[$$

*where $\Gamma^+$ and $\Gamma^-$ are the fundamental solutions of*

$$L^+ = \mu\Delta_{p_0} + \langle x, BD\rangle - \partial_t, \qquad L^- = \frac{1}{\mu}\Delta_p + \langle x, BD\rangle - \partial_t,$$

*respectively, and $\Delta_q$ denotes the Laplacian operator in $\mathbf{R}^q$.*

**Dirichlet Problem for $L$.** In this section we study the first boundary value problem for the operator

$$L = \sum_{i,j=1}^{p_0} a_{ij}(z_0)\partial_{x_i x_j} + \langle x, BD\rangle - \partial_t$$

considered in Section 3. We first need to introduce some notation. Denoting by $\Omega$ an open bounded subset of $\mathbf{R}^{N+1}$, we define

$$d(z) = \inf_{\zeta \in \partial_\Omega} \|\zeta^{-1} \circ z\|_B, \quad d(z, \bar{z}) = \min\{d(z), d(\bar{z})\}.$$

We say that a function $u$ belongs to $C_d^\alpha(\Omega)$ if

$$|u|_{d,\alpha} \stackrel{\text{def}}{=} \sup_\Omega |u| + \sup_{z,\bar{z} \in \Omega} (d(z, \bar{z}))^\alpha \frac{|u(z) - u(\bar{z})|}{\|\bar{z}^{-1} \circ z\|_B} < \infty$$

Moreover, $u \in C_d^{2+\alpha}(\Omega)$ if

$$|u|_{d,2+\alpha} \stackrel{\text{def}}{=} \sup_\Omega |u| + \sum_{i=1}^{p_0} \sup_\Omega d(z)|u_{x_i}(z)| + |d^2 Yu|_\alpha + \sum_{i,j=1}^{p} |d^2 u_{x_i,x_j}|_\alpha$$

where

$$|d^2 v|_\alpha = \sup_\Omega |d^2 v| + \sup_{z,\bar{z} \in \Omega} (d(z, \bar{z}))^{2+\alpha} \frac{|v(z) - v(\bar{z})|}{\|\bar{z}^{-1} \circ z\|_B}$$

In [M], Manfredini proved the following a priori Hölder estimates.

**Theorem 4.1.** *Suppose the coefficients $a_{i,j}$ and the function $f$ in the class $C_d^\alpha(\Omega)$ for a fixed positive $\alpha$, $\alpha < \frac{2}{2r+1}$ ($r$ is the number of the non zero blocks of $B$; see 3.2). Then if $u \in C_d^{2+\alpha}(\Omega)$ is a solution of the equation*

$$Lu = f$$

*then*

$$|u|_{2+\alpha,d} \leq c\{\sup_\Omega |u| + |d^2 f|_{\alpha,d}\}$$

*where the positive constant $c$ only depends on the Hölder norms of $a_{i,j}$ and on the ellipticity constant of the matrix $(a_{i,j})$.*

Starting from this result and using a continuity method, in [M] an existence theorem is proved for the first boundary value problem for $L$ on the particular class of the strongly regular sets, where an open subset $V$ of $\mathbf{R}^{N+1}$ is called *strongly regular* if

$$\forall z \in \partial V, \exists \quad \text{euclidean ball} \quad B(z + \nu, |\nu|) \subset \mathbf{R}^{N+1} \setminus V$$

**Theorem 4.2.** *Let the hypotheses of Theorem 4.1 be satisfied. Then, if $V$ is a bounded strongly regular open subset of $\mathbf{R}^{N+1}$, there exists a unique solution $u \in C_d^{2+\alpha}(V) \cap C(\bar{V})$ of the boundary problem*

$$\begin{cases} Lu = f & \text{in } V \\ u \mid_{\partial V} = 0 \end{cases}$$

The following corollary readily follows from the theorem above, the a-priori estimate and the maximum principle.

**Corollary 4.3.** *Let $V$ be a bounded strongly regular open subset of $\mathbf{R}^{N+1}$. Then, for every $\phi \in C(\partial V)$ there exists a solution $u \in C_d^{2+\alpha}(V) \cap C(\bar{V})$ of the Dirichlet problem*

$$\begin{cases} Lu = 0 & \text{in } V \\ u\,|_{\partial V} = \phi \end{cases}$$

The Corollary 4.3 and the a-priori estimates contained in Theorem 4.1 are the starting point for proving an existence theorem for the Dirichlet problem for $L$ on an *arbitrary* open subset $\Omega$ of $\mathbf{R}^{N+1}$. To this end, the prove mostly relies on the method of Perron, which is based on the following definition of $L$-super solution.

**Definition 4.4.** *Let $\Omega$ be an open subset of $\mathbf{R}^{N+1}$. A lower semi-continuous function $u : \Omega \to \mathbf{R}$ is called $L$-super-harmonic if, for every strongly regular open set $V$ such that $\bar{V} \subseteq \Omega$, we have*

$$u \geq H_\phi^V \quad \text{in} \quad V$$

*for every continuous function $\phi : \partial\Omega \to \mathbf{R}$, $\phi \leq u$. $H_\phi^V$ denotes the solution of the Dirichlet problem*

$$\begin{cases} Lu = 0 & \text{in } V \\ u\,|_{\partial V} = \phi \end{cases}$$

*which existence was proved in Corollary 4.3.*

It is worth noticing that the class $\{V\}$ of the strongly regular set is a basis for the euclidean topology of $\mathbf{R}^{N+1}$.

**Theorem 4.5.** *Let $\Omega$ be any bounded open set of $\mathbf{R}^{N+1}$. Let $\phi$ be a continuous function defined on $\partial\Omega$. Then there exists a generalized solution (in the sense of Perron) of the Dirichlet problem*

$$\begin{cases} Lu = 0 & \text{in } \Omega \\ u\,|_{\partial\Omega} = \phi & \phi \in C(\partial\Omega) \end{cases}$$

*The solution is given by*

$$u = u_\phi = \inf\{v / \quad v \quad L\text{-super-harmonic in } \Omega, \quad \liminf_{\partial\Omega} v \geq \phi\}.$$

As a consequence of this approach, the generalized solution $u$ is a classical solution of the equation $Lu = 0$ in $\Omega$. However, in general, it does not take the value $\phi$ on the boundary.

*Definition 4.6. A boundary point $z_0 \in \partial\Omega$ is L-regular if $u_\phi(z) \to \phi(z_0)$ as $z \to z_0$ for any function $\phi \in C(\partial\Omega)$.*

It can be proved that the "Fichera boundary point" $\Sigma_3 \cup (\mathrm{int}_{\partial\Omega}\Sigma_2)$ are $L$-regular. Finally, in [M], the Dirichlet problem for non homogeneous equation

$$\begin{cases} Lu = f & \text{in } \Omega \\ u \mid_{\partial\Omega} = \phi \end{cases}$$

for arbitrary given function $f \in C_d^\alpha(\Omega)$, and $\phi \in C(\partial\Omega)$ is solved. First, if $\Gamma$ is the fundamental solution for $L$, it is proved that

$$w(z) = -\int_\Omega \Gamma(z,\zeta) f(\zeta) d\zeta$$

is a $C_{loc}^{2+\alpha}(\Omega)$ solution of $Lu = f$ in $\Omega$. A s consequence, letting $u = v + w$, where $v$ is the Perron solution of 4.1, then $u$ is a $C_{loc}^{2,\alpha}(\Omega)$ solution of $Lu = f$ which assumes the boundary data $\phi$ at any $L$-regular point of $\partial\Omega$.

**References**


[AS] Aronson, D.G. and Serrin, J., Local behavior of solution of quasilinear parabolic equations, *Arch. Rational Mech. Anal.* **25** (1967), pp. 81-122.

[C] Chandrasekhar, C., Stochastic problems in physics and astronomy, *Rev. Modern Phys.*, **15**, (1943), pp. 1-89.

[H] Hörmander, L., Hypoelliptic second order differential equations, *Acta Math.*, **119** (1967), pp. 147-171.

[I] Il'in, A.M., On a class of ultraparabolic equations, *Dokl. Akad. Nauk SSSR* **159** (1964), pp. 1214-1217;English trans., *Soviet Math. Dokl.* **5** (1964), pp. 1673-1676.

[K] Kuptsovn, L.P., Fundamental solutions for a class of second-order elliptic-parabolic equations, *Differentcial'nye Uravnenija* **8** (1972), pp. 1649-1660; English trans., *Differential Equations* **8** (1972), pp. 1269-1278.

[LP] Lanconelli, E. and Polidoro, S., On a class of hypoelliptic evolution operators, *Rend. Sem. Mat. Univ. Pol. Torino, Partial Diff. Eqs.* **52:1** (1994), pp. 29-63.

[M] Manfredini, M., work in progress.

[P1] Polidoro, S., On a class of ultraparabolic operators of Kolmogorov-Fokker-Plank type, *Rend. Sem. Mat. Catania*, (to appear).

[P2] ——, A global lower bound for the fundamental solution of Kolmogorov-Fokker-Plank equations. Preprint.

[S] Shiryayev, A.N. (Ed.), Selected works of A.N. Kolmogorov, Vol II; Math. and its Appl., Kluver Ac. Publ. Vol 26 (1986).

[So] Sonin, I.M., On a class of degenerate diffusion processes, *Teoriya Veroyatn i ee Primen*, **12** (1967), pp. 540-547, English trans. *Theory Probab. Appl.* **12** (1967), pp. 490-496.

[W] Weber, M., The fundamental solution of a degenerate partial differential equation of parabolic type, *Trans. Amer. Math. Soc.* **71** (1951), pp. 24-37.

E. Lanconelli
Dipartimento di Matematica
Università di Bologna
ITALY

P.-L. LIONS, B. PERTHAME, P.E. SOUGANIDIS

# Weak stability of isentropic gas dynamics for $\gamma = 5/3$

## I. Introduction

We consider the system of isentropic gas dynamics in Eulerian coordinates

$$(I.1) \qquad \begin{cases} \partial_t \rho + \partial_x(\rho u) = 0, & \text{in } \mathbb{R} \times (0,\infty), \\ \partial_t(\rho u) + \partial_x(\rho u^2 + p(\rho)) = 0, & \\ \rho u = \rho^0 u^0 \text{ and } \rho = \rho^0 \geq 0 & \text{on } \mathbb{R} \times \{0\}, \end{cases}$$

where, for $\gamma > 1$,

$$(I.2) \qquad p(\rho) = \kappa \rho^\gamma \quad \text{with } \kappa = \tfrac{\theta^2}{\gamma} \text{ and } \theta = \tfrac{\gamma - 1}{2} > 0.$$

The constant $\kappa$ is irrelevant but our choice simplifies further notations. We give a proof that bounded families of entropy solutions $(\rho_\varepsilon, u_\varepsilon)$ to (I.1) converge, extracting subsequences, strongly in $L^1_{\text{loc}}$ to an entropy solution of the system. We restrict our attention to the case $\gamma = 5/3$ where the arguments are simpler than for other values of $\gamma < 3$.

The issues of stability and existence, through the limit of an appropriate diffusive system, of solutions of (I.1), (I.2) for $\gamma \in (1,3)$ in both Lagrangian and Eulerian variables was studied in [LPS]. The proofs are rather long due to the technical difficulties introduced by the need to consider a large number of functional derivatives, their products and their averages, etc. On the other hand, the ideas in [LPS] are rather simple as we attempt to illustrate in the present note. Indeed, the choice of $\gamma = 5/3$ requires only two derivatives what this means will become transparent below. The calculations are then simple and allow for a direct presentation.

The existence and stability of solutions to the nonlinear hyperbolic system of gas dynamics are long standing questions. The program was initiated by Di Perna

[D1] and [D2], and later extended by Chen [C] and [LPT]. The paper [LPS] not only simplifies the arguments of [D1], [D2] and [C] but also closes the gap $(5/3, 3)$.

The outline of this paper is as follows. We describe the entropy inequalities and our main result in the second Section. In Section III, we give a (nearly) complete proof for the case $\gamma = 5/3$ ; some preliminary results from [LPS] are not proved again here.

## II. General facts and main result

The system (I.1) is hyperbolic and admits two Riemann invariants

$$z = u - \rho^\theta \quad \text{and} \quad w = u + \rho^\theta. \tag{II.1}$$

However, it is not strictly hyperbolic because of the vacuum state $\{\rho(t, x) = 0\}$. Indeed, it can be shown that entropy solutions, in a sense defined below, satisfy

$$\rho(t, x) \geq 0, \tag{II.2}$$

and

$$\min z^0(x) \leq z(t, x) \leq w(t, x) \leq \max w^0(x), \tag{II.3}$$

where $z^0$ and $w^0$ are the initial values of the Riemann invariants (see [LPT] for instance). These estimates imply that $u(t, x), \rho(t, x)$ are naturally bounded by the initial data. The question of passing to the limit in bounded families of solutions is therefore natural. However, the statement of the main result in that direction requires some further notations.

As usual for hyperbolic systems (see Lax [L]), we are interested in pairs entropy-entropy flux $(\eta, H)$. These are functions $\eta(\rho, u), H(\rho, u)$ such that $\eta(t, x) = \eta(\rho(t, x), u(t, x))$ and $H(t, x) = H(\rho(t, x), u(t, x))$ satisfy, for all smooth solutions $(\rho(t, x), u(t, x))$ of the system (I.1), the idendity

$$\partial_t \eta(t, x) + \partial_x H(t, x) = 0. \tag{II.4}$$

A family of such entropies is given in [LPT]. For $\gamma = 5/3$ these formula read

$$(II.5) \qquad \eta(\rho, u) = \int_{\mathbb{R}} g(\xi)\chi(\rho, \xi - u) d\xi,$$

$$(II.6) \qquad H(\rho, u) = \int_{\mathbb{R}} g(\xi)[\theta\xi + (1-\theta)u]\chi(\rho, \xi - u) d\xi,$$

and

$$(II.7) \qquad \chi(\rho, u) = (\rho^{2\theta} - u^2)_+ \ ,$$

where $\theta = 1/3$. Here we have set $x_+ = \max(x, 0)$.

Moreover, $\eta$ is convex in $\rho$, $\rho u$ if and only if $g$ is convex. These convex entropies are central because it is usual to define *entropy solutions* of the system (I.1) as those satisfying

$$(II.8) \qquad \partial_t \eta + \partial_x H \leq 0 \quad \text{in } \mathcal{D}'(\mathbb{R} \times (0, \infty))$$

for all the convex entropies of the family given by the representation formula (II.5).

We may now state the main result we wish to prove for $\gamma = 5/3$, but which holds for any $\gamma > 1$ (see [LPS], [LPT]).

**Theorem** ([LPS]). *From any bounded family of entropy solutions $(\rho_\varepsilon, u_\varepsilon)$ to the system of isentropic gas dynamics (I.1), we may extract a subsequence such that $(\rho_\varepsilon, \rho_\varepsilon u_\varepsilon)$ converge almost everywhere (and thus in $L^p_{\text{loc}}$ for all $1 \leq p < +\infty$) to an entropy solution to (I.1).* □

The main difficulty in this theorem is that no bound on derivatives of the solution are available except for the special value $\gamma = 3$ ([LPT]), where regularizing effects however follow from averaging Lemmas. In this context, the method of proof proposed by Di Perna consists in applying compensated compactness to all the entropy inequalities (II.8).

Let us describe this method. Following Murat [M] and Tartar [T], we may associate to the family $(\rho_\varepsilon, u_\varepsilon)_{\varepsilon > 0}$ a Young measure $d\nu_{t,x}(\rho, u)$. We denote by $\langle \ \ \rangle$ the integration with respect to this measure. For instance we set

$$(II.9) \qquad \langle \chi(\xi) \rangle = \int \chi(\rho, \xi - u) d\nu_{t,x}(\rho, u),$$

and we recall that $d\nu$ is a probability measure for each $(t, x)$. Since the time and space indices $(t, x)$ are just inessential parameters, we skip this dependency until the end of the analysis. Then, compensated compactness applied to (II.8) yields, for all $\xi, \xi'$ in $\mathbb{R}$,

$$(II.10)\quad 2[\langle\chi(\xi)u\rangle\langle\chi(\xi')\rangle - \langle\chi(\xi)\rangle\langle\chi(\xi')u\rangle] = (\xi - \xi')[\langle\chi(\xi)\chi(\xi')\rangle - \langle\chi(\xi)\rangle\langle\chi(\xi')\rangle].$$

This is justified in [LPT]. Here, we have used specifically $\gamma = 5/3$.

Then, the strong convergence of the sequence $(\rho_\varepsilon, u_\varepsilon)$ is equivalent ([M],[T]) to the fact that $d\nu$ is a Dirac mass in $\rho$ and $u$. We already can see a first difficulty arising from the non-strict hyperbolic of the system (I.1). When the density $\rho$ vanishes, $u$ is not defined and we cannot expect $d\nu_{t,x}(\rho, u)$ to be a Dirac mass. In fact, we prove in [LPS] that, for almost all $(t, x)$,

$$(II.11)\qquad d\nu_{t,x}(\rho, u) = \delta(\rho - \underline{\rho}(t, x)) \otimes \delta(u - \underline{u}(t, x)) \quad \text{with } \underline{\rho}(t, x) \neq 0,$$

or, when $\underline{\rho}(t, x)) = 0$,

$$(II.12)\qquad d\nu_{t,x}(\rho, u) = \delta(\rho) \otimes d\varphi_{t,x}(u)$$

for some probability measure $\varphi_{t,x}(\cdot)$. This is the result we develop in Section III.

The lack of strict hyperbolicity alone is not the crucial difficulty. Indeed, in [LPT], the case $\gamma \geq 3$ is treated rather easily ; in [JPP] the case of chromatography, a $N \times N$ non strictly hyperbolic system, is also treated rather easily. The difficulty to prove (II.11), (II.12) from (II.10), when $\gamma < 3$, comes from the lack of strict hyperbolicity combined with a bad sign in the variation of eigenvalues with respect to Riemann invariants for the system. Notice that the lack of strict hyperbolicity also results in a single family of entropies, given here by the formula (II.5). For strictly hyperbolic systems, the outer intervals can be used to define new entropy functions $\chi$ (see [JPP], [LPS]), thus giving new entropy relations which enrich the set of relations (II.10) deduced from compensated compactness.

## III. Proof of the Theorem for $\gamma = 5/3$

Following Section II, the proof of the Theorem is reduced to proving that the Young measure associated with the sequence $(\rho_\varepsilon, u_\varepsilon)$ satisfies (II.11), (II.12). The only information we can use is the compensated compactness equality (II.10). The proof contains several steps which aim to analyzing the singularities in $\langle\chi(\xi)\rangle$ and $\langle\chi(\xi)^2\rangle$. This analysis leads to the desired relation.

*First step. Simplification of the formula (II.10)*

**Lemma III.1.** *The set* $C = \{\xi \in \mathbb{R}, \langle\chi(\xi)\rangle > 0\}$ *is a bounded open interval* $]z_{\min}, w_{\max}[$. *Moreover,* $(\rho_0, u_0) = \left((\frac{w_{\max} - z_{\min}}{2})^{1/\theta}, \frac{z_{\min} + w_{\max}}{2}\right)$ *belongs to the support of* $d\nu(\rho, u)$. □

**Lemma III.2.** *Either* $d\nu$ *is supported in the vacuum state* $\{\rho = 0\}$, *or the following holds true for all* $\xi, \xi', \xi''$ *in* $C$

$$(III.1)\quad \begin{aligned}(\xi' - \xi'')\langle\chi(\xi'')\chi(\xi')\rangle\langle\chi(\xi)\rangle = \\ (\xi - \xi'')\langle\chi(\xi'')\chi(\xi)\rangle\langle\chi(\xi')\rangle + (\xi' - \xi)\langle\chi(\xi)\chi(\xi')\rangle\langle\chi(\xi'')\rangle.\end{aligned}$$

□

The proof of these two lemmas is given in [LPS]. There is no simplification for $\gamma = 5/3$. Thus we do not repeat the proof here.

*2nd step. Improved regularity of* $\langle\chi(v)\rangle$

At this level the value $\gamma = 5/3$ simplifies the analysis. Denoting $\chi', \chi''$... the derivatives of $\chi(\rho, \xi - u)$ with respect to the variable $\xi$, we have, in the distributional sense,

$$(III.2)\qquad \chi''(\xi) = 2\rho^\theta(\delta(\xi - z) + \delta(\xi - w)) + S(\rho, \xi - u)$$

where $\delta(\cdot)$ denotes the Dirac mass and $S$ is a BV function ($= -$ twice the indicator function of the set $\xi \in [z, w]$).

The heart of the proof consists in analyzing the singularities arising in the formula (III.1) from those of $\chi$ in (III.2). Since $\chi$ is the fundamental solution of a wave equation with variable coefficients, these singularities are issued from their propagation from the "initial time", i.e., here, $\rho = 0$.

**Lemma III.3.** *Either (II.12) holds true or we have for all $\xi$, $\xi' \in \mathcal{C}$,*

$$(III.3)\qquad \begin{cases} (\xi'-\xi)\langle\chi(\xi)\chi(\xi')\rangle\langle\chi''(\xi)\rangle = (\xi'-\xi)\langle\chi(\xi')\chi''(\xi)\rangle\langle\chi(\xi)\rangle \\ \quad +2[\langle\chi'(\xi)\chi(\xi)\rangle\langle\chi(\xi')\rangle - \langle\chi'(\xi)\chi(\xi')\rangle\langle\chi(\xi)\rangle]. \end{cases}$$

□

*Proof of Lemma III.3.* (II.12) corresponds in the Lemma III.2 to $d\nu$ supported by the vacuum state and (III.3) is deduced from (III.1) as follows. We differentiate it twice with respect to $\xi$, then we take $\xi' = \xi$. This yields (III.3) with $\xi''$ in place of $\xi'$. Notice a difficulty in sending $\xi'$ to $\xi$ in terms like

$$\langle\chi(\xi')\chi(\xi'')\rangle\;\langle\chi''(\xi)\rangle$$

because measures are involved. It is possible because $\chi(\xi')$ converges uniformly to $\chi(\xi)$. □

A first consequence of the formula (III.3) is the better regularity of $\langle\chi(\xi)\rangle$ than $\chi(\xi)$ itself. This proves for instance that the Young measure, if it is not a Dirac mass, is diffuse.

**Corollary III.4.** *Either (II.12) holds true or*

$$(III.4)\qquad \langle\chi''(\xi)\rangle \in BV_{\text{loc}}(\mathcal{C}).$$

□

*Proof of Corollary III.4.* Divide (III.3) by $(\xi'-\xi)$ and set $\xi'$ to $\xi$. We obtain, arguing as before to take care of the measures, that

$$(III.5)\qquad \begin{cases} \langle\chi^2(\xi)\rangle\langle\chi''(\xi)\rangle = \langle\chi''(\xi)\chi(\xi)\rangle\langle\chi(\xi)\rangle \\ \quad +2\langle\chi'(\xi)\chi(\xi)\rangle\langle\chi'(\xi)\rangle - 2\langle\chi'(\xi)^2\rangle\langle\chi(\xi)\rangle. \end{cases}$$

Since $\chi(\xi)$ vanishes at places where $\chi''$ has Dirac masses, we deduce from (III.2) that $\chi(\xi)\chi''(\xi)$, as well as all the other terms in the right handside of (III.5), belongs to $BV(\mathcal{C})$. This proves (III.4) because $\langle\chi^2\rangle$ is positive in $\mathcal{C}$ (but vanishes at its boundary points $w_{\max}, z_{\min}$). □

*3rd step. Reduction of the Young measure*

The last step of the proof is another formula similar to (III.3).

**Lemma III.5.** *Either (II.12) holds true or for all* $\xi,\ \xi' \in \mathcal{C}$,

$$(III.6)\qquad \begin{cases} 2\langle\chi(\xi)\rangle[\langle\chi'(\xi')\chi''(\xi)\rangle - \langle\chi''(\xi')\chi'(\xi)\rangle] \\ \quad = (\xi'-\xi)[\langle\chi(\xi)\chi''(\xi')\rangle\langle\chi''(\xi)\rangle - \langle\chi(\xi)\rangle\langle\chi''(\xi')\chi''(\xi)\rangle] \\ \qquad +2[\langle\chi(\xi)\chi'(\xi')\rangle\langle\chi''(\xi)\rangle - \langle\chi''(\xi')\rangle\langle\chi'(\xi)\chi(\xi')\rangle]. \end{cases}$$

*Proof of Lemma III.5.* Differentiate (III.3) twice with respect to $\xi'$. This is legitimate in $\mathcal{D}'(\mathbb{R}^2)$, since $\xi,\ \xi'$ are two independent variables. □

To conclude the proof of the main stability Theorem, let us take the formula (III.6) with $\xi' = \xi + h$ for $h > 0$ and substract the same formula with $\xi' = \xi - h$. We use the formula (III.2) for $\chi''$ and notice that

$$\chi'(\xi+h)\,\delta(\xi-w) = 0,$$

because $\chi'(\eta)$ vanishes for $\eta$ larger than $w$. Hence, we obtain from the left handside of (III.6)

$$(III.7)\qquad \begin{cases} 2\langle\chi(\xi)\rangle[\langle\chi'(\xi+h)\delta(\xi-z)\rangle + \langle\delta(\xi+h-w)\chi'(\xi)\rangle \\ \quad \langle\chi'(\xi+h)S(\xi)\rangle - \langle S(\xi+h)\chi'(\xi)\rangle \\ \qquad -\langle\chi'(\xi-h)S(\xi)\rangle + \langle S(\xi-h)\chi'(\xi)\rangle]. \end{cases}$$

Since $\chi'$ and $S$ are bounded, we have

$$\langle\chi'(\xi+h)S(\xi)\rangle - \langle\chi'(\xi-h)S(\xi)\rangle \to 0$$

as $h$ tends to 0, in $L^p(\mathcal{C})$ for all $p$, using Lebesgue's theorem. Therefore, as $h$ tends to zero, (III.7) reduces to

$$(III.8) \qquad 2\,\langle\chi'(\xi)\rangle\,[\langle\rho^\theta\delta(\xi-z)\rangle+\langle\rho^\theta\delta(\xi-w)\rangle].$$

Arguing similarly the right handside of (III.6) tends to zero in the same limit. Indeed, the only dangerous terms are the Dirac masses in

$$-h\,\langle\chi(\xi)\rangle\,[\langle\chi''(\xi+h)\chi''(\xi)\rangle-\langle\chi''(\xi-h)\chi''(\xi)\rangle].$$

But these terms vanish exactly (there is a difficulty here because we cannot take $\xi=\xi'$ in the distributional sense and to be rigorous we need to regularize one term, as is shown in [LPS]). Therefore, we find that the quantity in (III.8) vanishes.

This means that the only Dirac masses in (III.8) appear when $\langle\chi(\xi)\rangle=0$ i.e. at $z_{\min}, w_{\max}$. Therefore, either (II.12) holds, either the Young measure is reduced to a single Dirac mass at the point $(\underline{\rho},\underline{u})$ with $\underline{u}+\underline{\rho}^\theta=w_{\max}, \underline{u}-\underline{\rho}^\theta=z_{\min}$. This concludes the proof of the main theorem.

## References


[C] G. Q. Chen, The theory of compensated compactness and the system of Isentropic Gas Dynamics, preprint.

[D1] R. Di Perna, Convergence of the viscosity method for Isentropic Gas Dynamics, Comm. Math. Phys. **91** (1983), 1-30.

[D2] R. Di Perna, Convergence of approximate solutions to conservation laws, Arch. Rat. Mech. Anal. **82** (1983), 27-70.

[LPS] P.-L. Lions, B. Perthame and P.E. Souganidis, Existence of entropy solutions for the hyperbolic systems of isentropic gas dynamics in Eulerian and Lagrangian coordinates, preprint.

[LPT] P. -L. Lions, B. Perthame and E. Tadmor, Kinetic formulation of the Isentropic Gas Dynamics and p-systems, Comm. Math. Physics **163** (1994), 415–431.

[JPP] F. James, Y.–J. Peng and B. Perthame, Kinetic formulation of the chromatography and some other hyperbolic systems, J. Math. Pures et Appl., to appear.

[L] P. Lax, *Hyperbolic systems of conservation laws and the mathematical theory of shock waves*, CBMS-NSF **11** (1973).

[M] F. Murat, Compacité par compensation, Ann. Sc. Norm. Sup. Pisa **5** (1978), 489-507.

[T] L. Tartar, Compensated compactness and applications to partial differential equations, in *Research Notes in Mathematics, Nonlinear Analysis and Mechanics*: Herriot-Watt Symposium Vol. 4, ed. R.J. Knops, Pitman Press (1979).

P.-L. Lions
CEREMADE, URA CNRS 749, Université Paris IX-Dauphine, Place de Lattre de Tassigny, 75775 Paris cedex 16, France

B. Perthame
Laboratoire d'Analyse Numérique, URA CNRS 189, Université, Paris VI-Pierre et Marie Curie, BC 187, 4 place Jussieu, 75252 Paris cedex 05, France

P.E. Souganidis
Department of Mathematics, University of Wisconsin-Madison, Madison, WI 53706, USA

M. MARINO, A. MAUGERI

# Regularity results for the derivatives of solutions to nonlinear nonvariational parabolic systems

## 1. Introduction

Let $n$ and $N$ be two integers greater than 1, let $\Omega$ be an open bounded subset of $R^n$ with points $x = (x_1, x_2, \dots, x_n), x_i \in R$, let $T$ be a positive real number and $Q$ the cylinder $\Omega \times (-T, 0)$. If $u$ is a function from $Q$ into $R^N$, we set

$$D_i u = \frac{\partial u}{\partial x_i}, \ Du = (D_1 u, D_2 u, \dots, D_n u),$$

$$H(u) = \{D_i D_j u\} = \{D_{ij} u\}, \ i, j = 1, 2, \dots, n.$$

The letter $X$ will denote the generic point $(x, t) \in R^n_x \times R_t$, $p$ is the generic vector in $R^{nN}$, $p = (p_1, p_2, \dots, p_n)$, $p_i \in R^N$, and $\xi = \{\xi_{ij}\}$, $i, j = 1, 2, \dots, n, \xi_{ij} \in R^N$, stands for a point in $R^{n^2 N}$.

In the cylinder $Q$ we will consider the following nonlinear nonvariational parabolic system of second order

$$-a\left(X, u, Du, H(u)\right) + \frac{\partial u}{\partial t} = b(X, u, Du), \tag{1.1}$$

where $a(X, u, p, \xi)$ and $b(X, u, p)$ are vectors of $R^N$, measurable in $X$, continuous in $(u, p, \xi)$ and $(u, p)$ respectively, satisfying the following conditions:

$$a(X, u, p, 0) = 0; \tag{1.2}$$

(A) *there exist three positive constants $\alpha, \gamma$ and $\delta$, with $\gamma + \delta < 1$, such that, $\forall u \in R^N$, $\forall p \in R^{nN}$, $\forall \tau, \eta \in R^{n^2 N}$ and for almost every $X \in Q$*

$$\|\sum_{i=1}^{n} \tau_{ii} - \alpha\left[a(X, u, p, \tau + \eta) - a(X, u, p, \eta)\right]\|^2 \le \gamma \|\tau\|^2 + \delta \|\sum_{i=1}^{n} \tau_{ii}\|^2 \ (^1);$$

(1) Condition (A) has previously been introduced by S. Campanato in [6] to gen-

(1.3) *there exists a constant c such that*

$$\|b(X,u,p)\| \leq c(1+\|u\|+\|p\|), \text{ for a.e. } X \in Q, \forall u \in R^N, \forall p \in R^{nN}.$$

We will show that the spatial gradient, $Du$, of the solutions $u$, of class $W^2(Q,R^N) = L^2(-T,0,H^2(\Omega,R^N)) \cap H^1(-T,0,L^2(\Omega,R^N))$, to system (1.1) is partial Hölder continuous in $Q$, namely there exists a singular set $Q_0 \subset Q$, closed in $Q$, such that: meas $Q_0 = 0$ and

$$Du \in C^{0,\alpha}(Q \setminus Q_0, d, R^{nN}), \quad \forall \alpha < 1,$$

where the Hölder continuity is related to the parabolic metric

$$d(X,Y) = \max\{\|x-y\|, |t-\tau|^{\frac{1}{2}}\}, \quad X=(x,t), \quad Y=(y,\tau).$$

This result will be obtained through the well known properties of the spaces $\mathcal{L}^{p,\lambda}$, introduced by S. Campanato.
In order to apply this theory, we first establish a Caccioppoli's type inequality from which, by means of a suitable Poincaré's inequality and the Gehring - Giaquinta - G. Modica lemma, we derive a local $L^q$-regularity for the matrix $H(u)$.

Next, taking into account some algebraic lemmas, we get an integral estimate for the vectors $H(u)$ and $\frac{\partial u}{\partial t}$ that ensures the properties

$$H(u), \frac{\partial u}{\partial t} \in L^{2,n+2-2\varepsilon}, \qquad \forall \varepsilon \in (0,1),$$

and then, by virtue of Poincaré's inequality

(1.4) $$\int_{Q(X,\rho)} \sum_{i=1}^{n} \|D_i u - (D_i u)_{Q(X,\rho)}\|^2 dY \leq$$

---

eralize the well known Cordes condition.

In [18] (see also [19]) it was shown that, if the vector $a(X,u,p,\xi)$ satisfies conditions (1.2) and (A), then the operator $\xi \to a(X,u,p,\xi)$ is elliptic.
From condition (A), by assuming $\eta = 0$, it follows

$$\|a(X,u,p,\tau)\| \leq c\|\tau\|, \quad \text{for a.e. } X \in Q, \ \forall u \in R^N, \ \forall p \in R^{nN}, \ \forall \tau \in R^{n^2N}.$$

$$\leq c\rho^2 \int_{Q(X,\rho)} \left( \|H(u)\|^2 + \|\frac{\partial u}{\partial t}\|^2 \right) dY,$$

which is true $\forall u \in W^2(Q(X,\rho), R^N) = L^2(t-\rho^2, t, H^2(B(x,\rho), R^N)) \cap H^1(t-\rho^2, t, L^2(B(x,\rho), R^N))$ (see [10], Lemma 2.II)(2), also the property:

$$Du \in \mathcal{L}^{2,n+4-2\varepsilon}, \qquad \forall \varepsilon \in (0,1),$$

being $n+4-2\varepsilon > n+2$, this implies the $(1-\varepsilon)$-Hölder continuity of the spatial gradient $Du$.

The analogous problem for quasilinear nonvariational systems of second order has been studied in [9] (elliptic case) and [10] (parabolic case). Moreover, in [6] (see also [4], [5], and [7]) is also investigated the partial Hölder continuity for nonlinear nonvariational elliptic systems of second order.

Finally, we recall the definitions and the main properties of the Morrey-Campanato spaces $L^{2,\lambda}$ and $\mathcal{L}^{2,\lambda}$ with respect to the parabolic metric $d(X,Y)$.

Let $\lambda$ be a nonnegative real number and $Q = \Omega \times (-T, 0)$, where $\Omega$ is an open bounded subset of $R^n$ with Sobolev's "cone property" (3).

The symbol $L^{2,\lambda}(Q, d, R^k)$, $k$ integer $\geq 1$, denotes the space of all functions $u \in L^2(Q, R^k)$ such that

$$\sup_{\substack{X \in \bar{Q} \\ \rho > 0}} \{\rho^{-\lambda} \int_{Q \cap I(X,\rho)} \|u(Y)\|^2 dY\} < +\infty,$$

---

(2) If $X = (x,t) \in R^n_x \times R_t$, $Q(X,\rho) = B(x,\rho) \times (t-\rho^2, t)$, where $B(x,\rho)$ is the open ball in $R^n$ of radius $\rho$ centered at $x$.

$(D_i u)_{Q(X,\rho)}$ is the integral average of $D_i u$ on $Q(X,\rho)$ :

$$(D_i u)_{Q(X,\rho)} = \fint_{Q(X,\rho)} D_i u(Y) dY = \frac{1}{\text{meas} Q(X,\rho)} \int_{Q(X,\rho)} D_i u(Y) dY.$$

(3) This property ensures that $Q$ is of class $\mathcal{A}$ (with respect to the parabolic metric $d$), that is there exists a positive constant $K_Q$ such that

$$\text{meas } [Q \cap I(X,\rho)] \geq K_Q \rho^{n+2}, \ \forall X \in Q \text{ and } 0 < \rho \leq \text{ diam } Q.$$

where $I(X, \rho)$ is the open ball of center $X$ and radius $\rho$ with respect to the parabolic metric $d$.

$\mathcal{L}^{2,\lambda}(Q, d, R^k)$ stands for the space of all functions $u \in L^2(Q, R^k)$ such that

$$\sup_{\substack{X \in \bar{Q} \\ \rho > 0}} \{\rho^{-\lambda} \int_{Q \cap I(X,\rho)} \|u(Y) - u_{Q \cap I(X,\rho)}\|^2 dY\} < +\infty,$$

where $u_{Q \cap I(X,\rho)}$ is the integral average of $u$ on $Q \cap I(X, \rho)$.

Under the above assumptions on $Q$, it is well known that (see [11], Theorem 3.I) :

i) $\mathcal{L}^{2,\lambda}(Q, d, R^k)$ *is isomorphic to* $L^{2,\lambda}(Q, d, R^k)$, *if* $0 \leq \lambda < n + 2$;

ii) $\mathcal{L}^{2,\lambda}(Q, d, R^k)$ *is isomorphic to* $C^{0,\alpha}(\bar{Q}, d, R^k)$ *with* $\alpha = \frac{\lambda - (n+2)}{2}$, *if* $\lambda > n + 2$.

## 2. Local $L^q$-regularity of the matrix $H(u)$

Let us prove that, for a solution $u \in W^2(Q, R^N)$ to system (1.1), the following "Caccioppoli's inequality" holds

$$\int_{Q(X^0,\sigma)} (\|H(u)\|^2 + \|\frac{\partial u}{\partial t}\|^2) dX \leq \tag{2.1}$$

$$\leq c\sigma^{-2} \Big\{ \sigma^{-2} \int_{Q(X^0,2\sigma)} \|u - P_{Q(X^0,2\sigma)}\|^2 dX +$$

$$+ \int_{Q(X^0,2\sigma)} \|D(u - P_{Q(X^0,2\sigma)})\|^2 dX\} + c \int_{Q(X^0,2\sigma)} \|b(X, u, Du)\|^2 dX,$$

where $Q(X^0, 2\sigma) = B(x^0, 2\sigma) \times (t^0 - (2\sigma)^2, t^0) \subset\subset Q$, $X^0 = (x^0, t^0) \in R^n_x \times R_t$, $\sigma > 0$, and $B(x^0, 2\sigma) = \{x \in R^n : \|x - x^0\| < 2\sigma\}$, the constant $c$ does not depend on $\sigma$, and $P_{Q(X^0,2\sigma)}$ is the vector-polynomial in $x$, of degree $\leq 1$, such that

$$\int_{Q(X^0,2\sigma)} D^\alpha (u - P_{Q(X^0,2\sigma)}) dX = 0, \quad \forall \alpha : |\alpha| \leq 1.$$

Fix the cylinder $Q(X^0, 2\sigma) \subset\subset Q$, and choose two real functions $\vartheta(x)$ and $\rho(t)$ of class $C^\infty_0(R^n)$ and $C^\infty(R)$ respectively, satisfying the properties

$$0 \leq \vartheta \leq 1, \quad \vartheta = 1 \quad \text{in } B(x^0, \sigma), \quad \vartheta = 0 \quad \text{in } R^n \setminus B(x^0, 2\sigma), \tag{2.2}$$

$$|D^\alpha \vartheta| \leq c\sigma^{-|\alpha|} \quad \text{for all multi-indices } \alpha, \tag{2.3}$$

$$
(2.4) \qquad \begin{cases} 0 \le \rho \le 1, \quad \rho(t) = 1 \quad \text{for} \quad t \ge t^0 - \sigma^2, \\ \rho(t) = 0 \quad \text{for} \quad t \le t^0 - (2\sigma)^2, \\ |\rho'(t)| \le c\sigma^{-2}. \end{cases}
$$

Now (1.1) and (1.2) imply

$$
\vartheta\rho\Delta u - \alpha\vartheta\rho\frac{\partial u}{\partial t} =
$$

$$
= \vartheta\rho\{\Delta u - \alpha[a\,(X, u, Du, H(u)) - a(X, u, Du, 0)]\} - \alpha\vartheta\rho b(X, u, Du), \text{ in } Q,
$$

from which, by using condition (A) (with $\tau = H(u)$ and $\eta = 0$), we deduce

$$
(2.5) \qquad \vartheta\rho\|\Delta u - \alpha\frac{\partial u}{\partial t}\| \le
$$

$$
\le \vartheta\rho[\gamma\|H(u)\|^2 + \delta\|\Delta u\|^2]^{\frac{1}{2}} + \alpha\vartheta\rho\|b(X, u, Du)\|, \quad \text{in } Q.
$$

On the other hand, by setting $\mathcal{U}(X) = \vartheta(x)\rho(t)\left(u - P_{Q(X^0,2\sigma)}\right)$ one has

$$
(2.6) \qquad \begin{cases} \mathcal{U} \in L^2(t^0 - (2\sigma)^2, t^0, H^2 \cap H_0^1(B(x^0, 2\sigma), R^N))\cap \\ \qquad\qquad \cap\, H^1(t^0 - (2\sigma)^2, t^0, L^2(B(x^0, 2\sigma), R^N)), \\ \mathcal{U}(x, t^0 - (2\sigma)^2) = 0 \quad \text{in } B(x^0, 2\sigma), \\ \Delta\mathcal{U} = \vartheta\rho\Delta u + A, \\ H(\mathcal{U}) = \vartheta\rho H(u) + B, \\ \frac{\partial\mathcal{U}}{\partial t} = \vartheta\rho'(u - P_{Q(X^0,2\sigma)}) + \vartheta\rho\frac{\partial u}{\partial t}, \end{cases}
$$

where

$$
(2.7) \qquad \begin{cases} A = \Delta\vartheta\rho(u - P_{Q(X^0,2\sigma)}) + 2\rho\sum_i D_i\vartheta\cdot D_i(u - P_{Q(X^0,2\sigma)}), \\ B = \rho D_{ij}\vartheta\cdot(u - P_{Q(X^0,2\sigma)}) + \rho D_i\vartheta\cdot D_j(u - P_{Q(X^0,2\sigma)}) + \\ \qquad\qquad + \rho D_j\vartheta\cdot D_i(u - P_{Q(X^0,2\sigma)}). \end{cases}
$$

The first condition in (2.6) allow us to use in what follows the estimates

$$
(2.8) \qquad \int_{Q(X^0,2\sigma)} [\|H(\mathcal{U})\|^2 + \alpha^2\|\frac{\partial\mathcal{U}}{\partial t}\|^2]dX \le \int_{Q(X^0,2\sigma)} \|\Delta\mathcal{U} - \alpha\frac{\partial\mathcal{U}}{\partial t}\|^2 dX,
$$

$$
(2.9) \qquad \int_{Q(X^0,2\sigma)} \|\Delta\mathcal{U}\|^2 dX \le \int_{Q(X^0,2\sigma)} \|\Delta\mathcal{U} - \alpha\frac{\partial\mathcal{U}}{\partial t}\|^2 dX,
$$

obtained by S. Campanato in [8], by means of the C. Miranda - G. Talenti inequality (see [17], [20]):

$$\int_\Omega \|H(w)\|^2 dx \leq \int_\Omega \|\Delta w\|^2 dx,$$

which is true $\forall w \in H^2 \cap H_0^1(\Omega)$ whenever $\Omega$ is of class $C^2$ and convex.

Then owing to (2.6) and (2.5) we have

$$\text{(2.10)} \qquad \|\Delta \mathcal{U} - \alpha \frac{\partial \mathcal{U}}{\partial t}\| \leq \vartheta\rho\|\Delta u - \alpha \frac{\partial u}{\partial t}\| + \|A\| + \alpha\vartheta|\rho'|\|u - P_{Q(X^0,2\sigma)}\| \leq$$

$$\leq \vartheta\rho[\gamma\|H(u)\|^2 + \delta\|\Delta u\|^2]^{\frac{1}{2}} + \alpha\vartheta\rho\|b(X,u,Du)\| +$$

$$+\|A\| + \alpha\vartheta|\rho'|\|u - P_{Q(X^0,2\sigma)}\|,$$

from which, $\forall \varepsilon > 0$ and for a.e. $X \in Q$,

$$\text{(2.11)} \qquad \|\Delta \mathcal{U} - \alpha \frac{\partial \mathcal{U}}{\partial t}\|^2 \leq (1+\varepsilon)\vartheta^2\rho^2[\gamma\|H(u)\|^2 + \delta\|\Delta u\|^2] +$$

$$+c(\varepsilon,\alpha)\{\|b\|^2 + \|A\|^2 + (\rho')^2\|u - P_{Q(X^0,2\sigma)}\|^2\}.$$

Taking into account (2.6) once again, one can estimate (2.11) in the following way

$$\|\Delta \mathcal{U} - \alpha \frac{\partial \mathcal{U}}{\partial t}\|^2 \leq (1+\varepsilon)^2\gamma\|H(\mathcal{U})\|^2 + (1+\varepsilon)^2\delta\|\Delta \mathcal{U}\|^2 +$$

$$+c(\varepsilon,\alpha,\gamma,\delta)\{\|b\|^2 + \|A\|^2 + \|B\|^2 + (\rho')^2\|u - P_{Q(X^0,2\sigma)}\|^2\}.$$

Furthermore, integrating on $Q(X^0, 2\sigma)$ and using (2.8), (2.9), and (2.4), we reach:

$$\int_{Q(X^0,2\sigma)} \|\Delta \mathcal{U} - \alpha \frac{\partial \mathcal{U}}{\partial t}\|^2 dX \leq (1+\varepsilon)^2(\gamma+\delta) \int_{Q(X^0,2\sigma)} \|\Delta \mathcal{U} - \alpha \frac{\partial \mathcal{U}}{\partial t}\|^2 dX +$$

$$+c(\varepsilon,\alpha,\gamma,\delta)\left\{ \int_{Q(X^0,2\sigma)} \|b\|^2 dX + \int_{Q(X^0,2\sigma)} (\|A\|^2 + \|B\|^2) dX + \right.$$

$$\left. +\sigma^{-4} \int_{Q(X^0,2\sigma)} \|u - P_{Q(X^0,2\sigma)}\|^2 dX \right\}.$$

The assumption $\gamma + \delta < 1$ allows us to choose $\varepsilon$ in such a way that

$$\text{(2.12)} \qquad \int_{Q(X^0,2\sigma)} \|\Delta \mathcal{U} - \alpha \frac{\partial \mathcal{U}}{\partial t}\|^2 dX \leq c(\varepsilon,\alpha,\gamma,\delta)\left\{ \int_{Q(X^0,2\sigma)} \|b\|^2 dX + \right.$$

$$+\int_{Q(X^0,2\sigma)}(\|A\|^2+\|B\|^2)dX+\sigma^{-4}\int_{Q(X^0,2\sigma)}\|u-P_{Q(X^0,2\sigma)}\|^2dX\Bigg\}.$$

Consequently, by virtue of (2.8) and identities $H(\mathcal{U})=H(u)$ and $\frac{\partial\mathcal{U}}{\partial t}=\frac{\partial u}{\partial t}$ in $Q(X^0,\sigma)$,

$$\int_{Q(X^0,\sigma)}(\|H(u)\|^2+\alpha^2\|\frac{\partial u}{\partial t}\|^2)dX\leq \tag{2.13}$$

$$\leq c(\varepsilon,\alpha,\gamma,\delta)\Bigg\{\int_{Q(X^0,2\sigma)}\|b\|^2dX+$$

$$+\int_{Q(X^0,2\sigma)}(\|A\|^2+\|B\|^2)dX+\sigma^{-4}\int_{Q(X^0,2\sigma)}\|u-P_{Q(X^0,2\sigma)}\|^2dX\Bigg\}.$$

We can estimate the integral

$$\int_{Q(X^0,2\sigma)}(|A\|^2+\|B\|^2)dX,$$

which appears in the right hand side of (2.13), by means of (2.7) and (2.3), and we obtain

$$\int_{Q(X^0,2\sigma)}(\|A\|^2+\|B\|^2)dX\leq c\sigma^{-2}\Bigg\{\sigma^{-2}\int_{Q(X^0,2\sigma)}\|u-P_{Q(X^0,2\sigma)}\|^2dX+$$

$$+\int_{Q(X^0,2\sigma)}\|D(u-P_{Q(X^0,2\sigma)})\|^2dX\Bigg\}.$$

Finally, this inequality together with (2.13) produces the desidered estimate (2.1).

Now, taking into account that, for $u\in W^2(Q(X^0,2\sigma),R^N)$ and $\sigma\in(0,1)$, the following Poincarè - Sobolev estimate holds (see [16], Lemma 2.2):

$$\sigma^{-2}\int_{Q(X^0,2\sigma)}\|u-P_{Q(X^0,2\sigma)}\|^2dX+ \tag{2.14}$$

$$+\int_{Q(X^0,2\sigma)}\|D(u-P_{Q(X^0,2\sigma)})\|^2dX\leq$$

$$\leq c\left[\int_{Q(X^0,2\sigma)}(\|H(u)\|^{\frac{2(n+2)}{n+4}}+\|\frac{\partial u}{\partial t}\|^{\frac{2(n+2)}{n+4}})dX\right]^{\frac{n+4}{n+2}}$$

where $c$ does not depend on $\sigma$, we may estimate the right hand side of Caccioppoli's inequality by using (2.14), and we obtain the following

**Lemma 2.1.** *If $u \in W^2(Q, R^N)$ is a solution to system (1.1) and if assumptions (1.2), (1.3) and (A) hold, then, $\forall Q(X^0, 2\sigma) \subset\subset Q$, with $\sigma < 1$, one has*

$$(2.15) \qquad \fint_{Q(X^0,\sigma)} (\|H(u)\|^2 + \|\frac{\partial u}{\partial t}\|^2) dX \leq$$

$$\leq c \left[ \fint_{Q(X^0,2\sigma)} (\|H(u)\|^{\frac{2(n+2)}{n+4}} + \|\frac{\partial u}{\partial t}\|^{\frac{2(n+2)}{n+4}}) dX \right]^{\frac{n+4}{n+2}} + c \fint_{Q(X^0,2\sigma)} \|b\|^2 dX,$$

*where $c$ does not depend on $\sigma$.*

Now the local $L^q$-regularity of the matrix $H(u)$ will be derived with a standard technique, namely, making use of the following Gehring - Giaquinta - G. Modica lemma (see [15], Proposition 5.1 or [2], Cap. II, Lemma 10.I):

**Lemma 2.2.** *If $U$ and $G$ are nonnegative functions on $Q$ with*

$$U \in L^r(Q), \; G \in L^s(Q), \quad 1 < r < s,$$

*and if, $\forall Q(X^0, 2\sigma) \subset\subset Q$, with $\sigma < 1$, one has*

$$(2.16) \qquad \fint_{Q(X^0,\sigma)} U^r dX \leq c \left( \fint_{Q(X^0,2\sigma)} U dX \right)^r + c \fint_{Q(X^0,2\sigma)} G^r dX, \; c > 1,$$

*then there exists $\varepsilon \in (0, s-r]$ such that $U \in L^p_{\text{loc}}(Q)$, $\forall p \in [r, r+\varepsilon)$ and, $\forall Q(X^0, 2\sigma) \subset\subset Q$, with $\sigma < 1$, one has*

$$(2.17) \qquad \left( \fint_{Q(X^0,\sigma)} U^p dX \right)^{\frac{1}{p}} \leq K \left( \fint_{Q(X^0,2\sigma)} U^r dX \right)^{\frac{1}{r}} +$$

$$+ K \left( \fint_{Q(X^0,2\sigma)} G^p dX \right)^{\frac{1}{p}},$$

*where the constants $K$ and $\varepsilon$ depend only on $c, r, s$ and $n$.*

Indeed, if we set

$$U = \left( \|H(u)\| + \|\frac{\partial u}{\partial t}\| \right)^{\frac{2(n+2)}{n+4}}, \; G = \|b\|^{\frac{2(n+2)}{n+4}}, \; r = \frac{n+4}{n+2}, \; s = \frac{n+4}{n},$$

then the assumptions of Lemma 2.2 are fulfilled (in particular assumption (2.16) is a consequence of estimate (2.15) proved above). Thus, inequality (2.17) holds and consequently

$$H(u) \in L^q_{\text{loc}}(Q, R^{n^2 N}), \; \frac{\partial u}{\partial t} \in L^q_{\text{loc}}(Q, R^N), \quad \texttt{with} \quad q > 2.$$

Therefore we have

**Theorem 2.1.** *If $u \in W^2(Q, R^N)$ is a solution to system (1.1) and if assumptions (1.2), (1.3) and (A) hold, then there exists $\bar{q} \in (2, \frac{2(n+2)}{n}]$ such that, $\forall q \in [2, \bar{q})$,*

$$H(u) \in L^q_{\text{loc}}(Q, R^{n^2 N}),\ \frac{\partial u}{\partial t} \in L^q_{\text{loc}}(Q, R^N),$$

*and, $\forall Q(X^0, 2\sigma) \subset\subset Q$, with $\sigma < 1$, one has*

$$\left[ ⨍_{Q(X^0,\sigma)} (\|H(u)\|^q + \|\frac{\partial u}{\partial t}\|^q) dX \right]^{\frac{1}{q}} \le \tag{2.18}$$

$$\le c \left[ ⨍_{Q(X^0,2\sigma)} (\|H(u)\|^2 + \|\frac{\partial u}{\partial t}\|^2) dX \right]^{\frac{1}{2}} +$$

$$+c \left[ ⨍_{Q(X^0,2\sigma)} \|b(X, u(X), Du(X))\|^q dX \right]^{\frac{1}{q}},$$

*where $c$ does not depend on $\sigma$.*

This result, besides to have interest in itself, provides the global and partial Hölder continuity of the solutions $u$ to system (1.1) and of their spatial gradient $Du$. In fact, to achieve the Hölder continuity of the solutions $u$, we may use the following procedure: from estimate (2.18), making use of the Hölder inequality, we get an "interior fundamental estimate" for $H(u)$ and $\frac{\partial u}{\partial t}$, so that

$$H(u) \in L^{2,\lambda}_{\text{loc}}(Q, d, R^{n^2 N})$$

and then, by means of Poincaré's inequality,

$$u \in \mathcal{L}^{2,\lambda+4}_{\text{loc}}(Q, d, R^N),$$

from which, if $n$ is "small" ($n < \lambda + 2$), the Hölder continuity of $u$ in $Q$ follows.

A similar technique allows to derive from Theorem 2.1 partial Hölder continuity results for the spatial gradient $Du$ of the solutions $u$ to system (1.1) for any $n$. The next section will be devoted to this matter.

## 3. Partial Hölder continuity of the vector $Du$

Let $u \in W^2(Q, R^N)$ be a solution in $Q$ to system (1.1). We suppose that the vectors $a(X,u,p,\xi)$ and $b(X,u,p)$, besides the assumptions introduced in Section 1, satisfy the following ones:

(3.1) *$a(X,u,p,\xi)$ is continuous in $(X,u,p)$, of class $C^1$ in $\xi$, with derivatives $\frac{\partial a}{\partial \xi_{ij}}$ uniformly continuous and bounded in $Q \times R^N \times R^{nN} \times R^{n^2N}$;*

(B) *there exists a real function $\omega(t)$ defined for $t \geq 0$, continuous, bounded, concave, increasing, and with $\omega(0) = 0$, such that $\forall X, Y \in Q$, $\forall u, \nu \in R^N$, $\forall p, \bar{p} \in R^{nN}$ and $\forall \xi, \tau \in R^{n^2N}$*

$$\|a(X,u,p,\xi) - a(Y,\nu,\bar{p},\xi)\| \leq \omega(d^2(X,Y) + \|u-\nu\|^2 + \|p-\bar{p}\|^2) \cdot \|\xi\|,$$

$$\left\| \frac{\partial a(X,u,p,\xi)}{\partial \xi} - \frac{\partial a(X,u,p,\tau)}{\partial \xi} \right\| \leq \omega\Big(\|\xi - \tau\|^2\Big)\,(^4).$$

Now, let us fix the cylinder $Q(X^0,\sigma)$, with $\sigma < 1$ and $Q(X^0, 2\sigma) \subset\subset Q$, and let us rewrite system (1.1), considered in $Q(X^0,\sigma)$, in the following way:

$$-a(X^0, u_\sigma, (Du)_\sigma, H(u)) + \frac{\partial u}{\partial t} =$$

$$= [a(X,u,Du,H(u)) - a(X^0,u_\sigma,(Du)_\sigma,H(u))] + b(X,u,Du) =$$

$$= \mathcal{B}_1 + b(X,u,Du)\ (^5).$$

By setting

$$\frac{\partial \tilde{a}(X,u,p,\eta)}{\partial \xi_{ij}} = \left\{ \int_0^1 \frac{\partial a^h(X,u,p,t\eta)}{\partial \xi_{ij}^k} dt \right\}, h,k = 1,2,\dots,N,$$

the preceding system can be rewritten again as

$$-\sum_{ij=1}^{n} \frac{\partial \tilde{a}(X^0,u_\sigma,(Du)_\sigma,H(u))}{\partial \xi_{ij}} D_{ij}u + \frac{\partial u}{\partial t} =$$

---

(4) $\frac{\partial a(X,u,p,\eta)}{\partial \xi} = \left\{ \frac{\partial a(X,u,p,\eta)}{\partial \xi_{ij}} \right\}$, $i,j = 1,2,\dots,n$.

(5) $u_\sigma = u_{Q(X^0,\sigma)}$, $(Du)_\sigma = (Du)_{Q(X^0,\sigma)}$, $(H(u))_\sigma = (H(u))_{Q(X^0,\sigma)}$.

$$= \mathcal{B}_1 + b(X, u, Du),$$

which is equivalent to

$$(3.2) \qquad -\sum_{ij=1}^{n} A_{ij}^0 D_{ij} u + \frac{\partial u}{\partial t} = \mathcal{B}_2 + \mathcal{B}_1 + b(X, u, Du),$$

where

$$A_{ij}^0 = \frac{\partial \tilde{a}(X^0, u_\sigma, (Du)_\sigma, (H(u))_\sigma)}{\partial \xi_{ij}},$$

$$\mathcal{B}_2 = -\sum_{ij=1}^{n} \left[ A_{ij}^0 - \frac{\partial \tilde{a}(X^0, u_\sigma, (Du)_\sigma, H(u))}{\partial \xi_{ij}} \right] D_{ij} u.$$

In $Q(X^0, \sigma)$ we decompose $u$ in the sum $w + \nu$, where $w$ is the solution in $Q(X^0, \sigma)$ of the Cauchy - Dirichlet problem

$$(3.3) \qquad \begin{cases} w \in L^2(t^0 - \sigma^2, t^0, H^2 \cap H_0^1(B(x^0, \sigma), R^N)) \cap \\ H^1(t^0 - \sigma^2, t^0, L^2(B(x^0, \sigma), R^N)), \\ w(x, t^0 - \sigma^2) = 0 \text{ in } B(x^0, \sigma), \\ -\sum_{ij=1}^{n} A_{ij}^0 D_{ij} w + \frac{\partial w}{\partial t} = \mathcal{B}_2 + \mathcal{B}_1 , \end{cases}$$

while $\nu \in W^2(Q(X^0, \sigma), R^N)$ is a solution to the linear system

$$-\sum_{ij=1}^{n} A_{ij}^0 D_{ij} \nu + \frac{\partial \nu}{\partial t} = b(X, u, Du).$$

It is well known that the vector $u$ satisfies the following "fundamental estimate" (see [1])

$$\int_{Q(X^0, \tau\sigma)} (\|H(\nu)\|^2 + \|\frac{\partial \nu}{\partial t}\|^2) dX \le c\tau^{n+2} \int_{Q(X^0, \sigma)} (\|H(\nu)\|^2 + \|\frac{\partial \nu}{\partial t}\|^2) dX +$$

$$+ c \int_{Q(X^0, \sigma)} \|b(X, u, Du)\|^2 dX, \ \forall \tau \in (0, 1).$$

This inequality owing to the linear growth condition on $b$ and an algebraic lemma, can be rewritten in the form

$$(3.4) \qquad \int_{Q(X^0, \tau\sigma)} (\|H(\nu)\|^2 + \|\frac{\partial \nu}{\partial t}\|^2) dX \le$$

$$\leq c\tau^{n+2-\varepsilon}\int_{Q(X^0,\sigma)}(\|H(\nu)\|^2+\|\frac{\partial\nu}{\partial t}\|^2)dX+$$

$$+c\Phi(u,X^0,\sigma)(\tau^{n+2-\varepsilon}+\sigma^2),\ \forall\tau\in(0,1)\text{ and }\forall\varepsilon\in(0,n],$$

where

$$\Phi(u,X^0,\sigma)=\sigma^{n+2}+\int_{Q(X^0,\sigma)}(\|u\|^2+\|Du\|^2+\|H(u)\|^2+\|\frac{\partial u}{\partial t}\|^2)dX.$$

On the other hand, the Cauchy-Dirichlet problem (3.3) admits a unique solution that fulfils the estimate (see [12] and [13])

$$\int_{Q(X^0,\sigma)}(\|H(w)\|^2+\|\frac{\partial w}{\partial t}\|^2)dX\leq \tag{3.5}$$

$$\leq c\int_{Q(X^0,\sigma)}\|\mathcal{B}_1\|^2dX+c\int_{Q(X^0,\sigma)}\|\mathcal{B}_2\|^2dX.$$

The right hand side of (3.5) can be estimated, through standard technique, by means of assumptions (B), Theorem 2.1 and well known Poincaré's inequalities, obtaining the new estimate

$$\int_{Q(X^0,\sigma)}\|\mathcal{B}_1\|^2dX+\int_{Q(X^0,\sigma)}\|\mathcal{B}_2\|^2dX\leq \tag{3.6}$$

$$\leq c\Phi(u,X^0,2\sigma)\Big\{[\omega(c\sigma^{-n}\Phi(u,X^0,2\sigma))]^{1-\frac{2}{q}}+$$

$$+[\omega(\fint_{Q(X^0,2\sigma)}\|H(u)-(H(u))_{2\sigma}\|^2dX)]^{1-\frac{2}{q}}\Big\}.$$

From (3.5) and (3.6) we finally infer

$$\int_{Q(X^0,\sigma)}(\|H(w)\|^2+\|\frac{\partial w}{\partial t}\|^2)dX\leq \tag{3.7}$$

$$\leq c\Phi(u,X^0,2\sigma)\Big\{[\omega(c\sigma^{-n}\Phi(u,X^0,2\sigma))]^{1-\frac{2}{q}}+$$

$$+[\omega(\fint_{Q(X^0,2\sigma)}\|H(u)-(H(u))_{2\sigma}\|^2dX)]^{1-\frac{2}{q}}\Big\}.$$

Then, $\nu$ satisfies (3.4) and $w$ (3.7). Since $u=\nu+w$, we obtain, $\forall\tau\in(0,1),\forall\varepsilon\in(0,n]$,

$$\int_{Q(X^0,\tau\sigma)}(\|H(u)\|^2+\|\frac{\partial u}{\partial t}\|^2)dX\leq$$

$$\le c\Phi(u,X^0,2\sigma)\Big\{\tau^{n+2-\varepsilon}+\sigma^2+[\omega(c\sigma^{-n}\Phi(u,X^0,2\sigma))]^{1-\frac{2}{q}}+$$

$$+[\omega(\fint_{Q(X^0,2\sigma)}\|H(u)-(H(u))_{2\sigma}\|^2dX)]^{1-\frac{2}{q}}\Big\}$$

from which, by virtue of the estimate

$$\int_{Q(X^0,\tau\sigma)}(\|u\|^2+\|Du\|^2)dX\le$$

$$\le c\tau^{n+2}\int_{Q(X^0,\sigma)}(\|u\|^2+\|Du\|^2)dX+$$

$$+c\sigma^2\int_{Q(X^0,\sigma)}(\|Du\|^2+\|H(u)\|^2+\|\frac{\partial u}{\partial t}\|^2)dX,$$

it follows, $\forall\tau\in(0,1)$, $\forall\varepsilon\in(0,n]$

$$\Phi(u,X^0,\tau\sigma)\le c\Phi(u,X^0,2\sigma)\Big\{\tau^{n+2-\varepsilon}+\sigma^2+$$

$$+[\omega(c\sigma^{-n}\Phi(u,X^0,2\sigma))]^{1-\frac{2}{q}}+[\omega(\fint_{Q(X^0,2\sigma)}\|H(u)-(H(u))_{2\sigma}\|^2dX)]^{1-\frac{2}{q}}\Big\}.$$

This inequality is trivially true also for $\tau\in[1,2)$.

The above arguments show that *if* $u\in W^2(Q,R^N)$ *is a solution in* $Q$ *to system* (1.1) *and if assumptions* (1.2), (1.3), (3.1), (A) *and* (B) *are fulfilled, then,* $\forall Q(X^0,\sigma)\subset\subset Q$, *with* $\sigma<2$, $\forall\tau\in(0,1)$, *and* $\forall\varepsilon\in(0,n]$, *one has*

$$\Phi(u,X^0,\tau\sigma)\le A\Phi(u,X^0,\sigma)\Big\{\tau^{n+2-\varepsilon}+\sigma^2+ \tag{3.8}$$

$$+[\omega(c\sigma^{-n}\Phi(u,X^0,\sigma))]^{1-\frac{2}{q}}+[\omega(\fint_{Q(X^0,\sigma)}\|H(u)-(H(u))_{\sigma}\|^2dX)]^{1-\frac{2}{q}}\Big\},$$

*where* $q\in(2,\bar{q})$(6).

Due to (3.8), it is easy to see the partial Hölder continuity in $Q$ of the vector $Du$. The technique is very similar to that used in Theorem 5.1 of [3]; therefore, we only give the main ideas.

Setting

$$Q_1=\{X\in Q:\lim_{\sigma\to 0}{}''\fint_{Q(X,\sigma)}\|H(u)-(H(u))_{Q(X,\sigma)}\|^2dY>0\}$$

---

(6) $\bar{q}$ is the constant ($>2$) that appears in Theorem 2.1.

and

$$Q_2 = \{X \in Q : \lim_{\sigma \to 0}{}' \sigma^{-n} \Phi(u, X, \sigma) > 0\},$$

one has

$$\text{meas } Q_1 = 0$$

as well as (see [14])

$$\mathcal{H}_n(Q_2) = 0,$$

where $\mathcal{H}_n$ is the $n$-dimensional Hausdorff measure with respect to the parabolic metric $d(X, Y)$.

Moreover, the following theorem holds.

**Theorem 3.1.** *If $u \in W^2(Q, R^N)$ is a solution to system (1.1) and assumptions (1.2), (1.3), (3.1), (A), and (B) are fulfilled, then, $\forall \varepsilon \in (0,1)$, it is possible to associate to each $X^0 \in Q \setminus (Q_1 \cup Q_2)$ a cylinder $Q(X^0, R_{X^0}) \subset Q \setminus Q_2$ and a positive number $\sigma_\varepsilon$ such that*

$$\Phi(u, Y, \tau\sigma_\varepsilon) \leq (1 + A)\tau^{n+2-2\varepsilon}\Phi(u, Y, \sigma_\varepsilon), \forall \tau \in (0,1), \forall Y \in Q(X^0, R_{X^0}).$$

*Moreover,*

$$\begin{aligned} H(u) &\in L^{2,n+2-2\varepsilon}(Q(X^0, R_{X^0}), R^{n^2 N}), \\ \frac{\partial u}{\partial t} &\in L^{2,n+2-2\varepsilon}(Q(X^0, R_{X^0}), R^N), \\ Du &\in \mathcal{L}^{2,n+4-2\varepsilon}(Q(X^0, R_{X^0}), R^{nN}). \end{aligned}$$

*Consequently, there exists a set $Q_0$, closed in $Q$, with*

$$Q_2 \subset Q_0 \subset Q_1 \cup Q_2\ ,$$

*such that*

$$Du \in C^{0,\alpha}(Q \setminus Q_0, R^{nN}),\ \forall \alpha < 1.$$

**Remark 3.1.** Theorem 3.1 provides the partial Hölder continuity in $Q$ of the spatial gradient of the solutions $u$ to system (1.1), under assumptions (1.2), (1.3), (3.1), (A), and (B).

In particular, assumption (1.3), concerning the linear growth of $b(X, u, p)$, could be generalized by replacing the constant 1 by a function $f(X)$ of class $L^2(Q)$.

It remains open the problem to achieve the partial Hölder continuity in $Q$ of the vector $Du$, under the assumption that $b(X,u,p)$, instead of the linear growth, has a "controlled growth", that is, $b(X,u,p)$ satisfies the condition

$$\|b(X,u,p)\| \leq c(f(X) + \|u\|^{\beta} + \|p\|^{\beta'}),$$

where $f(X) \in L^2(Q)$, and $\beta, \beta'$ must be chosen in such a way that if $u \in W^2(Q, R^N)$ then

$$b(X,u,Du) \in L^2(Q, R^N).$$

## References


[1] S. Campanato, *Equazioni paraboliche del secondo ordine e spazi $\mathcal{L}^{2,\vartheta}(\Omega,\delta)$*, Ann. Mat. Pura Appl., (4) 73 (1966), 55-102

[2] S. Campanato, *Sistemi ellittici in forma divergenza. Regolarità all'interno*, Quaderni Scuola Norm. Sup. Pisa, 1980

[3] S. Campanato, *Hölder continuity and partial Hölder continuity results for $H^{1,q}$-solutions of non-linear elliptic systems with controlled growth*, Rend. Sem. Mat. Fis. Milano, 52 (1982), 435-472

[4] S. Campanato, *A Cordes Type Condition for non Linear non Variational Systems*, Rend. Acc. Naz. dei XL, 13-fasc. 20 (1989), 307-321

[5] S. Campanato, *Non variational differential systems. A condition for local existence and uniqueness*, Proceedings of the Caccioppoli Conference, 1989, Ricerche di Matem., Suppl. vol. 40 (1991), 129-140

[6] S. Campanato, *$\mathcal{L}^{2,\lambda}$ Theory for non Linear non Variational Differential Systems*, Rend. Matem., (7) 10 (1990) (scritti matematici dedicati a M.A. Sneider), 531-549

[7] S. Campanato, *Non variational basic elliptic systems of second order*, Rend. Sem. Mat. Fis. Milano, 60 (1990), 113-131

[8] S. Campanato, *Non variational basic parabolic systems of secon order*, Rend. Mat. Acc. Lincei, (9) 2 (1991), 129-136

[9] S. Campanato, P. Cannarsa, *Second Order Nonvariational Elliptic Systems*, Boll. Un. Mat. Ital., (5) 17-B (1980), 1365-1394

[10] P. Cannarsa, *Second Order Nonvariational Parabolic Systems*, Boll. Un. Mat. Ital., (5) 18-C (1981), 291-315

[11] G. Da Prato, *Spazi* $\mathcal{L}^{(p,\vartheta)}(\Omega,\delta)$ *e loro proprietà*, Ann. Mat. Pura Appl., (4) 69 (1965), 383-392

[12] E. Gagliardo, *Problema al contorno per equazioni differenziali lineari di tipo parabolico in n variabili*, Ricerche di Matem., 5 (1956), 169-205

[13] E. Gagliardo, *Teoremi di esistenza e di unicità per problemi al contorno relativi ad equazioni paraboliche lineari e quasi lineari in n variabili*, Ricerche di Matem., 5 (1956),239-257

[14] M. Giaquinta, E. Giusti, *Partial regularity for the solutions to non-linear parabolic systems*, Ann. Mat. Pura Appl., (4) 97 (1973), 253-266

[15] M. Giaquinta, G. Modica, *Regularity results for some classes of higher order nonlinear elliptic systems*, J. Reine Angew. Math., 311/312 (1979), 145-169

[16] M. Marino, A. Maugeri, *Second order non linear non variational parabolic systems*, Rend. Matem., (7) 13 (1993), 499-527

[17] C. Miranda, *Su di una particolare equazione ellittica del 2° ordine a coefficienti discontinui*, Anal. Sti. Univ. <<Al. I. Cura>> Jasi Sert. Ib. Fiz., 11 (1965), 209-215

[18] D.K. Palagachev, *Global strong solvability of Dirichlet problem for a class of nonlinear elliptic equations in the plane*, Le Matematiche, 48 (1993), 311-321

[19] D.K. Palagachev, *On a class of second order nonlinear elliptic equations*, to appear on Compt. Rend. Bulg. Acad. Sci

[20] G. Talenti, *Sopra una classe di equazioni ellittiche a coefficienti misurabili*, Ann. Mat. Pura Appl., 69 (1965), 285-304

Mario Marino - Antonino Maugeri
Dipart. di Matematica,
Viale A. Doria, 6
95125 Catania - ITALY

G. TALENTI

# On complex-valued solutions to a 2D eikonal equation

## 1. Introduction.

In the present paper we outline forthcoming results by R.Magnanini and the author about *complex-valued* solutions $w$ to the following first-order partial differential equation

$$w_x^2 + w_y^2 + 1 = 0. \tag{1.1}$$

Equation (1.1) arises, e.g., in questions about characteristic surfaces of Laplace's equation raised by H.Lewy, and in the theory of diffraction by J. Keller and D. Ludwig.

Suppose

$$w = u + iv; \tag{1.2a}$$

i.e., suppose

$$u = \Re(w) \quad \text{and} \quad v = \Im(w), \tag{1.2b}$$

the real and imaginary parts of $w$. Equation (1.1) is equivalent to the following first-order system of partial differential equations

$$u_x^2 + u_y^2 + 1 = v_x^2 + v_y^2, \tag{1.3a}$$

$$u_x v_x + u_y v_y = 0 \tag{1.3b}$$

— in alternative notations,

$$|\nabla u|^2 + 1 = |\nabla v|^2,$$

$$\nabla u \cdot \nabla v = 0.$$

Equation (1.3a) implies that the length of the gradient of $v$ exceeds 1, and equals 1 exactly at the critical points of $u$. Equation (1.3b) tells us that the gradients of

$u$ and $v$ are orthogonal — thus the level lines of $u$ are lines of steepest descent of $v$ and the lines of steepest descent of $u$ are level lines of $v$.

System (1.3) can be easily decoupled. In fact, algebraic manipulations show that (1.3) can be recast either in the following form

$$\begin{bmatrix} v_x \\ v_y \end{bmatrix} = \pm\sqrt{1 + \frac{1}{u_x^2 + u_y^2}} \begin{bmatrix} -u_y \\ u_x \end{bmatrix}, \tag{1.4a}$$

or in the form of the following pair

$$\begin{bmatrix} u_x \\ u_y \end{bmatrix} = \pm\sqrt{1 - \frac{1}{v_x^2 + v_y^2}} \begin{bmatrix} v_y \\ -v_x \end{bmatrix}, \quad v_x^2 + v_y^2 \geq 1. \tag{1.4b}$$

Loosely speaking, equations (1.4a) make the gradient of $v$ available if and only if $u$ obeys the following partial differential equation

$$\frac{\partial}{\partial x}\left\{\sqrt{1 + \frac{1}{u_x^2 + u_y^2}}\, u_x\right\} + \frac{\partial}{\partial y}\left\{\sqrt{1 + \frac{1}{u_x^2 + u_y^2}}\, u_y\right\} = 0; \tag{1.5a}$$

the equations appearing in (1.4b) make the gradient of $u$ available if and only if $v$ obeys the constraint involved and the following partial differential equation

$$\frac{\partial}{\partial x}\left\{\sqrt{1 - \frac{1}{v_x^2 + v_y^2}}\, v_x\right\} + \frac{\partial}{\partial y}\left\{\sqrt{1 - \frac{1}{v_x^2 + v_y^2}}\, v_y\right\} = 0. \tag{1.6a}$$

Thus, system (1.3) is satisfied if and only if either $u$ satisfies equation (1.5a) and $v$ is given by (1.4a), or $v$ satisfies $v_x^2 + v_y^2 \geq 1$ and equation (1.6a) and $u$ is given by (1.4b).

Formally, equation (1.5a) is the Euler equation of the functional $J$ defined by

$$J(u) = \int_G j(|\nabla u|)dxdy, \;\; j(\rho) = \frac{\rho}{2}\sqrt{1+\rho^2} + \frac{1}{2} ln\left(\rho + \sqrt{1+\rho^2}\right)$$

— however, $J$ is *not* differentiable at any function $u$ having critical points.

Equations (1.5a) and (1.6a) can be recast in the following form

$$\{|\nabla u|^4 + u_y^2\}u_{xx} - 2u_x u_y u_{xy} + \{|\nabla u|^4 + u_x^2\}u_{yy} = 0, \tag{1.5b}$$

and

$$\{|\nabla v|^4 - v_y^2\}v_{xx} + 2v_xv_yv_{xy} + \{|\nabla v|^4 - v_x^2\}v_{yy} = 0, \tag{1.6b}$$

respectively — i.e., in the form of semilinear second-order partial differential e-quations with polynomial nonlinearities.

If the coefficients of $u_{xx}, u_{xy}, u_{yy}$ appearing on the left-hand side of (1.5b) are denoted by $a, 2b, c$, then

$$\begin{vmatrix} a & b \\ b & c \end{vmatrix} = |\nabla u|^6(|\nabla u|^2 + 1) \geq 0.$$

Hence equation (1.5b) should be qualified *elliptic* or *elliptic-parabolic* — notice that degeneracies occur at the critical points of solutions.

If $a, 2b, c$ denote the coefficients of $v_{xx}, v_{xy}, v_{yy}$ appearing on the left-hand side of (1.6b), then

$$\begin{vmatrix} a & b \\ b & c \end{vmatrix} = |\nabla v|^6(|\nabla v|^2 - 1).$$

Hence solutions $v$ to (1.6b), such that $|\nabla v|^2 > 1$, are elliptic; any real-valued solution $v$ to the equation

$$|\nabla v|^2 = 1$$

— the standard equation of geometrical optics, which implies equation (1.6b) indeed — is a parabolic solution to (1.6b). Solutions to (1.6b) whose graph is a *developable surface* are hyperbolic. Recall from differential geometry that developable surfaces in the 3-dimensional euclidean space are the surfaces having zero Gauss curvature; they are planes, cylinders, cones, and the surfaces spanned by the tangent straight lines to some saddle curve — the edge of regression. As is easy to see, the solutions to (1.6b) whose graph is a developable surface are exactly the solutions to any equation of the following form

$$v_x^2 + v_y^2 - v_x cos\, C - v_y sin\, C = 0,$$

where $C$ is an arbitrary constant.

Observe that a set of terms, appearing on the left-hand side of equations (1.5b) and (1.6b), has a special geometric meaning. In fact, equations (1.5b) and (1.6b) read

$$|\nabla u|\Delta u + h = 0, \tag{1.5c}$$

and

$$|\nabla v|\Delta v - k = 0, \tag{1.6c}$$

respectively. Here

$$h = div\frac{\nabla u}{|\nabla u|} = \left(u_x^2 + u_y^2\right)^{-3/2}\left(u_y^2 u_{xx} - 2u_y u_x u_{xy} + u_x^2 u_{yy}\right), \tag{1.7a}$$

$$k = div\frac{\nabla v}{|\nabla v|}. \tag{1.8a}$$

Formulas from differential geometry tells us that the absolute value of $h$ at a point $(x, y)$, where the gradient of $u$ does not vanish, is the *curvature* at $(x, y)$ of the *level line* of $u$ crossing $(x, y)$; the absolute value of $k$ at a point $(x, y)$ is the *curvature* at $(x, y)$ of the *level line* of $v$ crossing $(x, y)$. Equations (1.4) yield

$$\pm h = div\frac{1}{|\nabla v|}\begin{bmatrix} v_y \\ -v_x \end{bmatrix} = \left(v_x^2 + v_y^2\right)^{-3/2}\left\{-v_x v_y\left(v_{xx} - v_{yy}\right) + \left(v_x^2 - v_y^2\right)v_{xy}\right\}, \tag{1.7b}$$

$$\pm k = div\frac{1}{|\nabla u|}\begin{bmatrix} -u_y \\ u_x \end{bmatrix}; \tag{1.8b}$$

hence the absolute value of $h$ at a point $(x, y)$ is also the curvature at $(x, y)$ of the line of *steepest descent* of $v$ crossing $(x, y)$, and the absolute value of $k$ at a point $(x, y)$ is also the curvature at $(x, y)$ of the line of *steepest descent* of $u$ crossing $(x, y)$.

One aim of the present paper is to point out that the solutions to equation (1.5) have no isolated critical point. Equation (1.5c) plainly shows a related property, namely that, if a solution is smooth and has a *smooth line of critical points,* then such a line must be *straight.*

## 2. Special solutions.

Solutions to equation (1.5), exhibiting special properties, are displayed below.

(i) *Radial* solutions to equation (1.5) — i.e., those invariant under the group of rotations about the origin — are given by

$$u(x,y) = \frac{C}{2} ln \frac{C + \sqrt{C^2 - x^2 - y^2}}{C - \sqrt{C^2 - x^2 - y^2}} - \sqrt{C^2 - x^2 - y^2}, \tag{2.1a}$$

and satisfy

$$|\nabla u(x,y)|^2 = C^2/(x^2 + y^2) - 1. \tag{2.1b}$$

Here $C$ is any real non-zero constant. Thus a radial solution to equation (1.5) is defined in a disk where $x^2 + y^2 \leq C^2$, and the circle where $x^2 + y^2 = C^2$ is a *line of critical points.*

(ii) As is easy to check, $(x, y) \mapsto \frac{1}{i}\sqrt{x^2 + (y - i)^2}$ — a sort of distance of $(x, y)$ from $(0, i)$ — is a complex-valued solution to (1.1).The relevant real part is given by

$$u(x,y) = \frac{1}{\sqrt{2}} \left[ \sqrt{(x^2 + y^2 - 1)^2 + 4y^2} + 1 - x^2 - y^2 \right]^{1/2} \tag{2.2}$$

and satisfies equation (1.5). Observe that such a $u$ is infinitely differentiable in the plane cut along the rays where $y = 0$ and $|x| \geq 1$. Moreover $0 \leq u \leq 1$, and $u$ takes the value 1 at any point of the $y - axis$ — therefore, the straight line where $x = 0$ is *a locus of interior maximum points,* as well as *a locus of critical points.*

(iii) The machinery of Legendre transforms applies successfully to equation (1.5).

Recall that the *Legendre transform* of some function $u$ — a sufficiently smooth real-valued function of two real variables $x$ and $y$ — is the function $U$ whose value at any point $(p, q)$ from a subdomain of the euclidean plane equals the height above the origin of the tangent plane to the graph of $u$ which is othogonal to the direction $(p, q, -1)$. In formulas, $U$ is defined by the following set of equations

$$p = u_x(x,y),\ q = u_y(x,y),\ U(p,q) = xu_x(x,y) + yu_y(x,y) - u(x,y). \tag{2.3a}$$

Taking differentials gives $dU = xdp + ydq$, hence (2.3a) yield

$$x = U_p\ (p,q),\ y = U_q\ (p,q),\ u(x,y) = pU_p\ (p,q) + qU_q\ (p,q) - U(p,q) \tag{2.3b}$$

— the equations defining the inverse Legendre transform. Equations (2.3b) can be thought of as a parametric representation of the graph of $u$ : here the graph in question is viewed as the envelope of its tangent plane, and the parameters — $p$ and $q$ — define precisely the slope of the tangent plane. Equations (2.3a) and (2.3b) show that $(x,y) \mapsto (u_x(x,y), u_y(x,y))$ and $(p,q) \mapsto (U_p\ (p,q), U_q\ (p,q))$ — the *hodograph mappings* — are inverse of each other; hence the hessian matrices of $u$ and $U$ are inverse of each other, i.e.,

$$\begin{bmatrix} u_{xx} & u_{xy} \\ u_{xy} & u_{yy} \end{bmatrix} = \left(U_{pp}U_{qq} - U_{pq}^2\right)^{-1} \begin{bmatrix} U_{pp} & -U_{pq} \\ -U_{pq} & U_{qq} \end{bmatrix} \tag{2.3c}$$

— a formula which gives the second-order derivatives of $u$ in terms of the second-order derivatives of $U$.

Formulas (2.3) tells us that $u$ satisfies equation (1.5) if and only if the Legendre transform $U$ of $u$ satisfies

$$\left((p^2+q^2)^2+p^2\right)\frac{\partial^2 U}{\partial p^2} + 2pq\frac{\partial^2 U}{\partial p\, \partial q} + \left((p^2+q^2)^2+q^2\right)\frac{\partial^2 U}{\partial q^2} = 0, \tag{2.4a}$$

a *linear* partial differential equation.

Equation (2.4a) reads

$$(\rho^2+1)\frac{\partial^2 U}{\partial \rho^2} + \rho\frac{\partial U}{\partial \rho} + \frac{\partial^2 U}{\partial \omega^2} = 0, \tag{2.4b}$$

if *polar coordinates* are introduced in the *hodograph plane,* i.e.,

$$p = \rho cos\omega,\ q = \rho sin\omega. \tag{2.5a}$$

Separation of variables and Fourier series arguments show that the solutions $U$ to equation (2.4) that are smooth in a punctured neighborhood of the origin are given by

$$\begin{aligned} U(\rho e^{i\omega}) =& A_0 + B_0 ln\left(\sqrt{1+\rho^2}+\ \rho\ \right) + \\ & \sum_{n=0}^{\infty}\left[A_n\left(\sqrt{1+\rho^2}+\ \rho\ \right)^n + B_n\left(\sqrt{1+\rho^2}-\ \rho\ \right)^n\right]cos(n\omega) + \\ & \sum_{n=0}^{\infty}\left[C_n\left(\sqrt{1+\rho^2}+\ \rho\ \right)^n + D_n\left(\sqrt{1+\rho^2}-\ \rho\ \right)^n\right]sin(n\omega), \end{aligned} \tag{2.5b}$$

where $A_n, B_n, C_n, D_n$ are constants. Formulas (2.3b) and (2.5a) give

$$\begin{aligned} x &= \cos\omega \frac{\partial U}{\partial \rho}(\rho e^{i\omega}) - \frac{\sin\omega}{\rho}\frac{\partial U}{\partial \omega}(\rho e^{i\omega}) \\ y &= \sin\omega \frac{\partial U}{\partial \rho}(\rho e^{i\omega}) + \frac{\cos\omega}{\rho}\frac{\partial U}{\partial \omega}(\rho e^{i\omega}) \\ u(x,y) &= \rho\frac{\partial U}{\partial \rho}(\rho e^{i\omega}) - U(\rho e^{i\omega}). \end{aligned} \tag{2.5c}$$

Formulas (2.5) produce solutions to equation (1.5). For instance, selecting appropriate constants gives $U(\rho e^{i\omega}) = \rho\sqrt{1+\rho^2}\cos(2\omega)$, hence (2.5) give

$$\begin{aligned} +x &= (1+\rho^2)^{-1/2}\cos\omega\big[2(1+\rho^2) - \cos(2\omega)\big] \\ -y &= (1+\rho^2)^{-1/2}\sin\omega\big[2(1+\rho^2) + \cos(2\omega)\big] \\ u(x,y) &= \rho^3(1+\rho^2)^{-1/2}\cos(2\omega). \end{aligned} \tag{2.6}$$

It can be shown that equations (2.6) define a genuine — i.e., single-valued — solution to equation (1.5); such a solution is defined in the exterior of the curve where

$$(x^2+y^2-4)^3 + 27(x^2-y^2)^2 = 0$$

— an asteroid — and this asteroid is a *locus of critical points.*

## 3. Critical points.

The specific solutions displayed in section 2 have no isolated critical points. The following theorem indicates that all solutions to equation (1.5b) have the same property.

**Theorem 2.** Let $w$ be a smooth solution to equation (1.1), and let $u$ be the real part of $w$. Then $u$ *cannot* have isolated critical points.

Proof. Let

$$\boldsymbol{l} = \frac{1}{|\nabla u|}\begin{bmatrix} -u_y \\ u_x \end{bmatrix}, \tag{3.1a}$$

a unit vector field whose trajectories are the level lines of $u$ and whose divergence is — in absolute value — the curvature of the lines of steepest descent of $u$. Equation (3.1a) gives

$$\begin{bmatrix} 0 & 1 \\ -1 & 0 \end{bmatrix}\boldsymbol{l} = \frac{\nabla u}{|\nabla u|}, \tag{3.1b}$$

a unit vector field whose trajectories are the lines of steepest descent of $u$ and whose divergence is — in absolute value — the curvature of the level lines of $u$.

Crucially, equations (1.4) and (3.1) imply that $\boldsymbol{l}$ is *smooth everywhere,* even across the critical points of $u$.

As a consequence, the level lines of $u$ are free from singular points, and the lines of steepest descent have a smooth curvature. Since

$$\nabla|\nabla u| = \begin{bmatrix} u_{xx} & u_{xy} \\ u_{xy} & u_{yy} \end{bmatrix}\frac{\nabla u}{|\nabla u|}, \tag{3.2}$$

we infer also that $|\nabla u|$ is continuously differentiable everywhere, even near the critical points of $u$.

An inspection shows that

$$\frac{\partial}{\partial \boldsymbol{l}}|\nabla u| + (div\, \boldsymbol{l})|\nabla u| = 0. \tag{3.3}$$

In conclusion, if $u$ has a critical point — the origin, say — then the critical points of $u$ must spread along the level line of $u$ which crosses the origin. □

Giorgio Talenti
Università di Firenze, Dipartimento di Matematica,
viale Morgagni 67A, 50134 Firenze, Italy.
e-mail address: talenti@udini.math.unifi.it

N.S. TRUDINGER

# Isoperimetric inequalities and nonlinear elliptic equations

In this talk we describe the application of isoperimetric inequalities, recently obtained by us in [8], to estimates for Hessian operators. The estimates hold for associated admissible functions and extend previous estimates of Tso [13] for convex functions. Details of proofs are provided in [12].

## 1. Hessian operators

We consider operators of the form

$$F_k[u] = [D^2u]_k\,, \qquad k = 1, \cdots, n\,, \tag{1}$$

where $u \in C^2(\Omega)$, $\Omega$ is a domain in Euclidean $n$-space $\mathbb{R}^n$, and for any real $n \times n$ matrix $\mathcal{A}$, $[\mathcal{A}]_k$ denotes the sum of the $k \times k$ principal minors of $\mathcal{A}$. We may also write

$$F_k[u] = S_k(\lambda[D^2u])\,, \tag{2}$$

where $\lambda = \lambda[D^2u]$ denotes the eigenvalues of the Hessian matrix $D^2u$, and $S_k$ is the $k$-th elementary symmetric function, given by

$$S_k(\lambda) = \sum_{1 \le i_1 < \cdots < i_k \le n} \lambda_{i_1} \cdots \lambda_{i_k}\,. \tag{3}$$

A function $u \in C^2(\Omega)$ is *admissible* with respect to the operator $F_k$, or simply *k-admissible*, if

$$[D^2u + \eta]_k \ge [D^2u]_k \tag{4}$$

for all non-negative real $n \times n$ symmetric matrices $\eta$. When $k > 1$, the function $u$ is $k$-admissible if and only if $F_j[u] \ge 0$, $j = 1, \cdots, k$. The operator $F_k$ is degenerate elliptic with respect to admissible functions. Note that for $k = 1$, $F_1[u] = \Delta u$ and all functions $u \in C^2(\Omega)$ are admissible, while for $k = n$, $F_n[u] = \det D^2u$ and the admissible functions are those which are locally convex in $\Omega$.

For $k = 0, 1, \cdots, n-1$, we define the domain $\Omega$ to be *k-convex*, (*uniformly k-convex*), if $\partial\Omega \in C^2$ and

$$H_j[\partial\Omega] \ge 0\,,\ (>0)\,, \quad j = 0, \cdots, k\,,$$

where $H_j[\partial\Omega]$ denotes the $j^{th}$ mean curvature of $\partial\Omega$, given by

$$H_j[\partial\Omega] = S_j(\kappa)\,, \qquad (S_0 \equiv 1)\,,$$

where $\kappa = (\kappa_1, \cdots, \kappa_{n-1})$ are the principal curvatures of $\partial\Omega$, (oriented so that $\kappa_i \geq 0$ for convex $\Omega$). Note that $\Omega$ is $(n-1)$-convex if and only if its components are convex. If $u$ is admissible with respect to the operator $F_k$, then the non-degenerate level surfaces of $u$ are $(k-1)$-convex. In the other direction we have the following existence result of Caffarelli, Nirenberg and Spruck [4].

**Theorem 1** *Assume* $\partial\Omega \in C^\infty$, $\phi \in C^\infty(\partial\Omega)$, $\psi \in C^\infty(\bar{\Omega})$, $\inf_\Omega \psi > 0$ *and* $\Omega$ *is uniformly* $(k-1)$*-convex for some* $k = 1, \cdots, n$*. Then there exists a unique admissible solution* $u \in C^\infty(\bar{\Omega})$ *of the Dirichlet problem:*

$$F_k[u] = \psi \quad in \quad \Omega\,, \qquad u = \phi \quad on \quad \partial\Omega\,. \tag{5}$$

**Remark.** If only $\psi \geq 0$ in $\Omega$, $\psi^{1/k} \in C^{1,1}(\bar{\Omega})$, we obtain a unique admissible solution $u \in C^{1,1}(\bar{\Omega})$, (Krylov [5]).

## 2. Quermassintegrals

For $k = 0, \cdots, n$, the quermassintegral $V_k(\Omega)$ was introduced by Minkowski as a mean $k$-dimensional cross-sectional area of the *convex* domain $\Omega$. For arbitrary domains $\Omega$ in $I\!R^n$, with boundary $\partial\Omega \in C^2$, we adopt the definitions

$$\begin{aligned} V_k(\Omega) &= \frac{1}{n\binom{n-1}{k}} \int_{\partial\Omega} H_{n-k-1}[\partial\Omega]\,, \qquad k = 0, \cdots, n-1\,, \\ V_n(\Omega) &= |\Omega|\,. \end{aligned} \tag{6}$$

For a ball $B_R$ of radius $R$, we clearly have $V_k(B_R) = \omega_n R^k$ while for $k = n-1$ we obtain

$$V_{n-1}(\Omega) = \frac{1}{n} \times \text{ perimeter of } \Omega\,.$$

$V_0$ is proportional to the Euler-Poincaré characteristic of $\Omega$. Quermassintegrals are fundamental to integral geometry [6]. Provided the domain $\Omega$ is $(n-k-1)$-convex, whenever $1 \leq k \leq n-2$, the quermassintegral $V_k(\Omega) \geq 0$, and we can define the $k$-mean radius of $\Omega$, $\zeta_k$, by

$$\zeta_k(\Omega) = \left[V_k(\Omega)/\omega_n\right]^{1/k}\,. \tag{7}$$

The isoperimetric inequalities in [8] may then be simply expressed as:

**Theorem 2** *For $1 \leq l \leq k \leq n$, we have $\zeta_k \leq \zeta_l$.*

**Remarks.** For *convex* $\Omega$, these are consequences of the famous Aleksandrov-Fenchel inequalities for mixed volumes, [23]. When $k = n$, $l = n - 1$, we obtain the classical isoperimetric inequality with no geometric restrictions. Equality holds if and only if $\Omega$ is a ball [8].

**Ingredients of proof.** Our proof in [8] rests upon the solution and gradient estimates of Hessian equations of the form

$$\det D^2u = \psi[D^2u]_k \,, \qquad k = 0, \cdots, n-1 \,, \tag{8}$$

where $\psi$ approximates the characteristic function of $\Omega$, together with integration formulae

$$\int_\Omega [D^2u]_k = \int_{\partial\Omega} \int_0^{\gamma\cdot Du} [t\partial\gamma + \partial^2 u]_{k-1}\, dt \,, \quad k = 1, \cdots, n \,, \tag{9}$$

where $\gamma$ denotes the outer unit normal vector field on $\partial\Omega$ and $\partial$ is the tangential gradient operator in $\partial\Omega$. The classical solvability of the above equations is treated in [10].

## 3. Symmetrization

Let $\Omega$ be a bounded $(k-1)$-convex domain in $I\!R^n$, $k = 1, \cdots, n$, and let $u \in C^\infty(\Omega) \cap C^0(\bar{\Omega})$ be $k$-admissible in $\Omega$ with $u = 0$ on $\partial\Omega$. For almost all $t \in (m, 0]$, $m = \inf_\Omega u$, the domaïn

$$\Omega_t = \{x \in \Omega \,|\, u(x) < t\}$$

will have boundary $\sum_t \in C^\infty$ given by the level surface

$$\sum_t = \{x \in \Omega \,|\, u(x) = t\} \,.$$

Accordingly we define the $(k-1)$-symmetrand of $u$, by

$$u^*_{k-1}(x) = \sup \{t \,|\, \zeta_{n-k+1}(\Omega_t) < |x| \,,\ Du \neq 0 \text{ on } \sum_t\} \,. \tag{10}$$

**Properties** [12]. *Setting $\rho(r) = u^*(x)$ for $r = |x|$, we have that $\rho$ is a non-decreasing function in $C^{0,1}[0, R]$ with $\rho(0) = m$, $\rho(R) = 0$, where $R = \zeta_{n-k+1}(\Omega)$.*

Returning to the Dirichlet problem (5), we let $\psi^0$ denote the symmetric rearrangement of $\psi$ and consider the radially symmetric problem:

$$F_k[u^0] = \psi^0 \text{ in } B_R, \quad u^0 = 0 \text{ on } \partial B_R. \tag{11}$$

By Tso [13], problem (11) has the explicit solution

$$u^0(x) = -[n/\binom{n}{k}]^{1/k} \int_{|x|}^{R} \left[s^{k-n} \int_0^s \psi^0(t)\, t^{n-1}\, dt\right]^{1/k} ds, \tag{12}$$

where we have written $\psi^0(t) = \psi^0(x)$, $|x| = t$.

**Theorem 3** [12] *Let $u$ be the solution to the Dirichlet problem* (5). *Then*

$$\min_{\partial\Omega} \phi + u^0 \leq u^*_{k-1} \leq \max_{\partial\Omega} \phi \tag{13}$$

*in $B_R$.*

From Theorem 3, follow various estimates for the solution of (5). For example, we obtain estimates in $L^\infty(\Omega)$ if $\psi \in L^q(\Omega)$ for $(k+1)\,q > n$. The estimate (13) will also apply to weak (or viscosity) solutions of (5), when the smoothness of the data is reduced.

## 4. Sobolev type inequalities

Again suppose that $u \in C^\infty(\Omega) \cap C^0(\bar{\Omega})$ is $k$-admissible in $\Omega$ and vanishes in $\partial\Omega$. Writing

$$F_k^{ij}(\mathcal{A}) = \frac{\partial}{\partial a_{ij}} [\mathcal{A}]_k, \quad \mathcal{A} = [a_{ij}],$$

we then have

$$\left[F_k^{ij}(D^2 u)\right] \geq 0 \quad \text{in} \quad \Omega.$$

We define the integral functionals,

$$I_{k,p}[u;\Omega] = \int_\Omega F_k^{ij}\, D_i u D_j u\, |Du|^{p-k-1} \tag{14}$$

for $p \geq 1$. When $p \leq k-1$, the integrand in (14) is assumed to vanish when $Du = 0$.

**Theorem 4** [12]

$$I_{k,p}[u;\Omega] \geq I_{k,p}[u^*_{k-1}; B_R].$$

As a corollary of Theorem 4 we obtain sharp inequalities of Sobolev type.

**Theorem 5** [12] *For $k+p<n+1$, $q=np/(n+1-k-p)$ we have*

$$\|u\|^p_{L^q(\Omega)} \le C(n,k,p)\, I_{k,p}[u;\Omega]\,, \tag{15}$$

*with optimal constants*

$$C(n,k,1) = [k\binom{n}{k}\,{\omega_n}^{k/n}]^{-1} \tag{16}$$

*and $C(n,k,p)$ for $p>1$, determined by the radial functions*

$$u(x) = [a+|x|^{(k+p-1)/(p-1)}]^{(k+p-n-1)/(k+p-1)}\,, \quad a>0\,, \tag{17}$$

*with $\Omega = I\!R^n$.*

The case $p=1$ in Theorem 5 follows directly from the isoperimetric inequality, Theorem 2, with $k=n$, $l=n-k$. The cases $p>1$ depend on the symmetrization result, Theorem 4. Note that in the case $p=k+1$, we can write

$$\begin{aligned} I_{k,k+1}[u;\Omega] &= \int_\Omega F^{ij}\, D_i u\, D_j u \\ &= -k\int_\Omega u F_k[u]\,, \end{aligned} \tag{18}$$

with associated Euler-Lagrange operator

$$\delta\, I_{k,k+1}[u;\Omega] = \frac{k}{k+1}\, F_k[u]\,. \tag{19}$$

Furthermore the quantities $(I_{k,k+1})^{1/(k+1)}$ furnish a norm [14] and the resultant imbeddings into $L_{q'}$ for $q'<q$ in Theorem 5 are compact. As a consequence, we are able to treat equations of the form

$$F_k[u] + g(\kappa, u) = 0\,, \tag{20}$$

yielding extensions of well known theory for the case $k=1$.

Finally, we remark that the condition that $u$ is admissible with respect to the Hessian operator $F_k$ in Theorems 3, 4 and 5 can generally be weakened to the condition

$$F_k^{ij}\, D_i u\, D_j u \ge 0\,. \tag{21}$$

In particular, Theorems 4 and 5 will also apply to admissible functions for the $k$-mean curvature operator [11].

## References


[1] C. Bandle, *Isoperimetric Inequalities and Applications*, Pitman, 1980.

[2] Yu.D. Burago and V.A. Zalgaller, *Geometric Inequalities*, Springer–Verlag, 1988.

[3] H. Busemann, *Convex Surfaces*, Wiley Interscience, 1958.

[4] L. Caffarelli, L. Nirenberg and J. Spruck, The Dirichlet problem for nonlinear and second order elliptic equations, III: Functions of the eigenvalues of the Hessian, *Acta Math.* **155** (1985), 261–301.

[5] N.V. Krylov, *Lectures on Fully Nonlinear Second Order Elliptic Equations*, Univ. of Bonn, 1994.

[6] L. Santalo, *Integral Geometry and Geometric Probability*, Addison–Wesley, 1977.

[7] G. Talenti, Some estimates of solutions to Monge-Ampère equations in dimension two, *Ann. Mat. Scuola Norm. Sup. Pisa* **8** (1981), 183–230.

[8] N.S. Trudinger, Isoperimetric inequalities for quermassintegrals, *Ann. Inst. Henri Poincaré — Analyse non linéaire*, **11** (1994), 411–425.

[9] N.S. Trudinger, Isoperimetric inequalities and Monge-Ampère equations. In: *A Tribute to Ilya Bakelman, Discourses in Mathematics and Its Applications*, Texas A&M University, No. 3 (1994), 43–46.

[10] N.S. Trudinger, On the Dirichlet problem for Hessian equations, *Acta Math.* (1995).

[11] N.S. Trudinger, Maximum principles for curvature quotient equations, *J. Math. Sci. Univ. of Tokyo* 1 (1994), 398-413.

[12] N.S. Trudinger, On new isoperimetric inequalities and symmetrization (in preparation.

[13] K. Tso, On symmetrization and Hessian equations, *J. d'Analyse Math.* **52** (1989), 94–106.

[14] X-J. Wang, A class of fully nonlinear elliptic equations and related functionals. *Indiana Univ. Math. J.* **43** (1994), 25–44.

Neil Trudinger

Centre for Mathematics and its Applications

The Australian National University

Canberra

A.C.T. 0200